ABOUT ISLAND PRESS

Island Press, a nonprofit organization, publishes, markets, and distributes the most advanced thinking on the conservation of our natural resources—books about soil, land, water, forests, wildlife, and hazardous and toxic wastes. These books are practical tools used by public officials, business and industry leaders, natural resource managers, and concerned citizens working to solve both local and global resource problems.

Founded in 1978, Island Press reorganized in 1984 to meet the increasing demand for substantive books on all resource-related issues. Island Press publishes and distributes under its own imprint and offers these services to other nonprofit organizations.

Support for Island Press is provided by Geraldine R. Dodge Foundation, The Energy Foundation, The Charles Engelhard Foundation, The Ford Foundation, Glen Eagles Foundation, The George Gund Foundation, The William and Flora Hewlett Foundation, The John D. and Catherine T. MacArthur Foundation, The Andrew W. Mellon Foundation, The Joyce Mertz-Gilmore Foundation, The New-Land Foundation, The J. N. Pew, Jr., Charitable Trust, Alida Rockefeller, The Rockefeller Brothers Fund, The Rockefeller Foundation, The Tides Foundation, and individual donors.

ABOUT INTERNATIONAL TROPICAL TIMBER ORGANIZATION

Established under an international treaty as a forum for consultation and coooperation, International Tropical Timber Organization commenced operations at its headquarters in Yokohama, Japan, in 1987. With a membership comprising the world's major producer and consumer countries of tropical timber, ITTO is dedicated to promoting the conservation and wise utilization of tropical forest resources such that by the year 2000, tropical timber products traded internationally will be sourced from sustainably managed forests.

NOT BY TIMBER ALONE

NOT BY TIMBER ALONE

*Economics
and Ecology
for Sustaining
Tropical Forests*

Theodore Panayotou
and Peter S. Ashton

ISLAND PRESS

Washington, D. C. ☐ *Covelo, California*

This book is based on a study commissioned by the International Tropical Timber Organization (ITTO) to the Harvard Institute of International Development (HIID). The support of ITTO is gratefully acknowledged. The views expressed, however, are those of the authors and do not necessarily represent the views or policy of either ITTO or HIID.

Library of Congress Cataloging-in-Publication Data

Panayotou, Theodore.
    Not by timber alone : economics and ecology for sustaining
tropical forests / Theodore Panayotou and Peter S. Ashton.
        p.   cm.
    Includes bibliographical references (p.   ) and index.
    ISBN 1-55963-195-3 (acid-free paper).—ISBN 1-55963-196-1 (pbk.
    : acid-free paper)
        1. Forests and forestry—Tropics.   2. Forest products—Tropics.
    3. Forest products industry—Tropics.   4. Forests and forestry—
    Tropics—Multiple use.   5. Sustainable forestry—Tropics.
    6. Forest policy—Tropics.   7. Forest ecology—Tropics.   I. Ashton,
    Peter S.   II. Title.
    SD247.P68   1992
    333.75′0913—dc20   92-10222
                                                           CIP

Printed on recycled, acid-free paper

Manufactured in the United States of America

10   9   8   7   6   5   4   3   2   1

Contents

List of Tables

List of Figures

Foreword

Dr. B. C. Y. Freezailah

Perhaps only the world's oceans can rival forests in their capacity to generate such a diverse array of useful products, servicing the needs of many generations of users. The potential of moist tropical forests to discharge this function excels not only because of their well-known species richness, which varies dramatically among and within the great formations, but also because such numbers of people still dwell within the ecotone between the tropical forest and its cultivated environs, depending for much of their livelihood upon transmitted knowledge concerning the forest's multifarious produce.

Unique among commodity agreements for its emphasis on the conservation and sustainable management of tropical forests, the International Tropical Timber Organization (ITTO) has never been content simply to deliver bland statements on the much-debated themes of sustainability, biodiversity, and product diversification. Rather, it has recognized the dangers inherent in the unthinking use of ill-defined terms, and has sought to give them substance through various initiatives, in particular the convening of expert panels to frame guidelines for good forest management that may then serve as prototypes for its member governments, and the commissioning of research on specific issues that are at once topical and enduring.

Multiple-use management of tropical forests is undoubtedly one of these issues. One older Central European theory that has made later appearances on other continents, the so-called wake theory, argued that non-timber forest products (then known collectively as "minor") would be produced automatically, like the wake of a ship, when the helmsman set a course for the point on his binnacle marked "timber production." Such a theory is no longer tenable. Instead, thorough research is needed to establish multiple-objective production functions,

Dr. B. C. Y. Freezailah is the executive director of the International Tropical Timber Organization.

a task which must encompass an analysis of the undervaluation of tropical forests that sets the pace for today's deforestation, destroying timber and wildlife habitat alike, methods of valuing non-marketed and subsistence products, studies of new logging technologies with improved environmental sensitivity, and recommendations on incentives to encourage multiple-use management by the owners of tropical forest, still largely governments.

To this end, the ITTO commissioned a distinguished research organization, the Harvard Institute for International Development (HIID), to examine the case for multiple-use management of tropical hardwood forests, drawing upon the experience of ITTO's tropical timber-producing countries, now 22 in number, who together comprise over 80 percent of the world's closed tropical forest in Africa, Latin America, and the Caribbean, and the Asia/Pacific regions. The authors have brought within the covers of one volume a systematic study that reviews the condition of the resource base and the extent of the trade and a basic summary of the non-timber forest products and environmental services, such as water quality maintenance and soil conservation, which are also forest products, *sensu stricto*. They analyze the economics of multiple-use management and consider the changes needed in harvesting technology and plantation design to accommodate multiple-use management. A special section is devoted to the genetic resource; the problems of land tenure, executive institutions, government policies, and international cooperation are discussed in their appropriate contexts.

The HIID study concludes with an agenda for the future, and it is my pleasure to conclude on the same forward-looking note by recommending this publication to readers in the firm expectation that it will widen conceptual horizons, deepen vague ideas into sound thoughts leading to innovative actions, and lift hopes for the future capacity of the world's tropical forests to continue producing a wide range of products and services. For such outputs benefit those living in and around the forests, those farming lands downstream of the forest or processing its raw materials in other parts of the country, and those enjoying, in faraway places, either the forest material worked up into fine furniture, or just the "option demand" for the continued existence of the forest and the welfare of those whose livelihood depends upon it.

—Yokohama, March 1992

Foreword

DWIGHT H. PERKINS

There are few topics more important for the future of our planet than what happens over the next few decades to what remains of the world's tropical forests. Too often, however, the issue is posed as being a conflict between the requirements of growth and a higher standard of living on the one hand, and the conditions needed to maintain a sustainable environment on the other. Posing the problem in this way is a formula for ecological disaster. The people who live in and around the earth's great tropical forests are not going to give up their chance for a higher standard of living in order to reduce carbon dioxide in the atmosphere or to conserve biological diversity for the people of the world as a whole. Successful management of our environment necessitates finding paths to development that are consistent with a sustainable environment. The two must stand together or both will fall separately.

The integration of nature preservation and economic growth is central to the research and policy advisory work of the Harvard Institute for International Development (HIID). The very essence of sustainable development is the recognition of the fundamental interdependence between economic growth and natural resource management. Without efficient use and conservation of resources, there can be no lasting improvement in living standards. And without improvement in living standards in the developing world, there can be no conservation of resources in the long run. HIID is working with developing-country governments and research institutions on economic and social policy reforms to bring about sustained growth, poverty alleviation, and natural resource conservation.

In this context, the two authors of this volume pose the fundamental

Dwight H. Perkins is the director of the Harvard Institute for International Development.

choices not as growth versus the environment but as a question of how to make the sustainability of tropical forests economically as well as environmentally rewarding. In presenting their argument, they deal not only with the often inappropriate management and pricing of timber resources, but with a myriad of other forest products and services actual and potential. They look at these questions both from the perspective of the world's need to protect its genetic resource bank and the global climate and from the perspective of the local populations, who must survive physically and also advance economically if they are to reduce their dependence on the forest over time.

Studies of most policy issues do not fit neatly within a single academic discipline, and the study of policies to sustain tropical forests is no exception. This volume is coauthored by an economist and a botanist, but it also draws on work in other natural and social science disciplines. The objective is to understand and suggest remedies for a critical world problem. The approach is rigorously analytical in a multidisciplinary context.

In addition to their academic qualifications, both authors have extensive experience with their subject in the field. Combined, they have over three decades of residence in the tropics, which have been the focus of their professional activity, be it research, teaching, or policy advising. This study is itself a product of collaboration between the two authors that stretches over several years of shared interests and collaborative research.

—Cambridge, Massachusetts, March 1992

Acknowledgments

This book would not have been possible without the financial support, close cooperation, and assistance of the International Tropical Timber Organization (ITTO). In 1987, ITTO commissioned the Harvard Institute for International Development (HIID) to review the state of knowledge on multiple-use management of tropical hardwood forests and the potential role that non-timber forest products and services can play in the sustainment of the forests. The study, which was completed in early 1988, was revised and updated in 1990 and again in early 1992. The authors express their appreciation to ITTO and its member countries for their support and cooperation. We are especially thankful to the executive director of ITTO, Dr. B. C. Y. Freezailah, and to Nils Svanquist for their constant help and encouragement. We also acknowledge the support of HIID during the latest updating and editing, as well as the Rockefeller Foundation, the John Merck Fund, and the Alton Jones Foundation for partially funding administrative costs and authors' salaries during the revision.

In preparing the first draft of the study, we consulted many experts in the field. We thank them all, although it is possible to mention by name only a few: Dr. Philip R. O. Kio of the Forestry Research Institute of Nigeria; Dr. Gary Hartshorn of the Tropical Science Center, San José, Puerto Rico (now of World Wildlife Fund, United States); Dr. Alan Grainger of Resources for the Future (now at the University of Leeds); Dr. Jeffrey R. Vincent, the Department of Forestry, Michigan State University (now at HIID); Dr. Ricardo Godoy, HIID; Dr. Mark Leighton of the Department of Anthropology, Harvard University; and Dr. Charles Peters, New York Botanical Garden.

We also wish to acknowledge the input of several research assistants who helped us to review a voluminous literature and reviewed early drafts: Alexander Moad, Ph.D. candidate in biology, Harvard University; Gordon Foer, M.A. candidate in urban and environmental policy, Tufts University; Songpol Jetanavanich, Ph.D. candidate in economics, Boston University; Vesna Karaklejic, B.A. candidate in biology, Harvard University; Catherine A. Crumbley, Ph.D. candidate in environmental science, University of Massachusetts; Charles P. O'Hara, M.P.A. student, John F. Kennedy School of Government; and Mack Choi, Ph.D. candidate in the urban planning program, Harvard University. Finally, we thank Dr. James Ito-Adler, Mona Yacoubian, and Christopher L. Shaw, who edited the 1988, 1990, and 1992 versions of this study, respectively, as well as Alison Brucker, Scott Parsons, and Katherine Philp, who performed the word-processing of a difficult manuscript. Christopher Shaw also assisted with the updating and revising of the final draft. Special thanks go to our colleague, Jeffrey Vincent, who patiently read the manuscript and made valuable comments and suggestions for improvement. The authors remain, of course, solely responsible for the views expressed and any errors committed.

NOT BY
TIMBER
ALONE

CHAPTER 1

Introduction and Overview

THE CURRENT debate over the use of tropical forests pits economic growth directly against conservation. Exploiting the forests for hardwood timber is often considered the growth-oriented option, while production of non-timber goods and services is seen as environmentally sound but unprofitable. However, multiple-use management for timber and non-timber goods and services both maximizes economic growth and conserves the forest's value for the future. Indeed, the value of hardwood-producing forests can be enhanced with full accounting of non-timber goods and forest services through multiple-use management: Under certain conditions, in fact, the value of such non-timber products and services can surpass the value of the standing timber. Non-timber forest products are defined as non-timber wood products such as fuelwood, charcoal, pulp, chips for composite materials, fencing, poles, and implements, as well as non-wood forest goods such as game, fruits and nuts, cords and fibers, and latexes. Forest services include environmental benefits and ecological services derived from natural forests such as soil, water, and species conservation, recreation and tourism, protection against natural disasters such as floods and landslides, and preservation of wildlife and biological diversity. Tropical forests also serve as regulators of regional and global climate patterns through evapotranspiration and carbon sequestration.

This book has three objectives: first, to identify and evaluate the non-timber products and services that can be obtained from tropical forests; second, to determine the extent to which full accounting and enhancement of these products and services in a multiple-use management framework would help ensure the sustainability of tropical

3

hardwood timber supplies; and third, to identify gaps in knowledge and make recommendations for future research. The study covers the tropical forest regions of Africa, Latin America, and Asia/Pacific.

Since World War II, international trade in tropical timber products has increased enormously, while real prices have risen only modestly. The demand for tropical hardwoods is expected to continue rising as a result of population and income growth. Even if per capita consumption of forest products in developed countries declined, as it has in the United States, total domestic consumption in most tropical countries is expected to double in 25 to 35 years, given present population growth rates (2 to 4 percent) in these countries. This anticipated growth in domestic consumption will limit tropical timber exports and exert upward pressure on tropical timber prices. More threatening to the tropical timber trade, however, are the ongoing depletion and degradation of the resource base and the mounting opposition of indigenous populations and environmental groups to destructive logging practices in natural forests.

The U.S. Interagency Task Force on Tropical Forests concluded in 1980 that if present trends continue, the world's tropical forests outside Central Africa and the Amazon Basin would be "nothing but scattered remnants" by the year 2025. By the turn of the century, Latin America is expected to be supplying two-thirds of all exports of tropical hardwood timber (Grainger 1987). Prices will be substantially higher because harvesting and transport costs will increase as timber sources shift from Southeast Asia to the more heterogeneous and distant Amazonian forests, assuming that substitution with temperate timber is limited. The basic problem is not the anticipated higher prices but the continuing failure of these prices to reflect the increasing scarcity of the timber resource, the deteriorating condition of the resource base, and the inadequate investment response to the expected price increases.

UNDERVALUATION OF TIMBER RESOURCES

In economic terms, the most serious problem faced by the tropical timber trade is the undervaluation of the resource by governments of tropical countries. As owners of over 80 percent of the world's tropical forests, developing country governments have been unwilling or unable to capture more than a small fraction (10 to 50 percent) of the stumpage value or scarcity rent of the timber resource. The undervaluation of tropical hardwood timber and its resource base, the tropical high forest, combined with overvaluation of the net benefits from forest conversion, has led to excessive deforestation, failure to implement natural forest management, and underinvestment in forest plantations.

The problem is further compounded by these factors: (1) insecurity of tenure resulting from logging concessions that are shorter than felling cycles, lack of concession transferability, and uncertain renewability; (2) logging concessions that are awarded on a political rather than an economically competitive basis; (3) regulations that require concessionaires to begin harvesting their sites by a stipulated time; (4) tax structures based on marketable timber removed, rather than the potentially marketable timber on the site, thereby encouraging high-grading and damage to the remaining stand; and (5) disregard of customary use rights, which leads to interference and encroachment on concessions by members of local communities.

A second economic problem is the common failure to account for the non-timber forest products and services in forest management and investment decisions. Non-timber forest products are generally referred to as "minor" forest products and are treated as such in those few cases where they are not totally ignored. For local communities especially, they are of major economic importance. Some, such as rattan and latex, are also significant in international trade.

A list of non-timber forest products would include thousands of items: exudates (gums, resins, and latex); canes (rattan and bamboo); edible nuts, fruits, vegetables, and fungi; game animals and fish; flowers and fodder; and innumerable plants with biochemically active and useful substances, including those for medicinal and pharmaceutical uses, condiments, and spices. Of the most important of these commercial products, rattan and bamboo are found in Asia; wildlife is prominent in Africa; and fruits, nuts, and fish are common in Latin America. Exports of non-timber forest products from Indonesia reached $120 million in 1982, an amount almost half as large as the government revenues from timber production, notwithstanding the fact that non-timber exports were far smaller in volume than timber exports. In some areas, such as the Iquitos region of the Peruvian Amazon, the potential value of non-timber forest products was found to exceed that of hardwood timber (Peters et al. 1989).

The management of multi-species tropical forests for both timber and non-timber products is best considered from the perspective of ecological guilds of species sharing similar regeneration requirements, rather than from that of individual species with different requirements. The production of hardwood timber and non-timber forest goods need not be mutually exclusive; instead, they can be joint products arising from the complementary exploitation of the same environmental resource base. Ecologically and silviculturally, the two cannot be easily separated; optimal management for the one may or may not constitute optimal management for the other, but reconciliation may often be possible.

Economically, for instance, non-timber forest products can both increase the return from silvicultural improvements and plantation investments and help alleviate a major disadvantage of forest investments relative to alternative investments. Forest investments in tropical hardwoods generally involve a long gestation period (50 to 70 years) between expenditures and returns, which creates a serious cash-flow problem in the often imperfect and distorted capital markets of developing countries. Non-timber forest products can provide an annual income to alleviate this cash-flow problem, thus affording the critical margin necessary for forest investments to attract scarce capital and land from competing uses. Considering the value of non-timber forest products can make the difference between a socially acceptable and sustainable timber industry and a logging enclave resented by the local population.

ENVIRONMENTAL SERVICES

In addition to timber and non-timber products, tropical forests also provide important environmental benefits or services, such as regulation of droughts and floods, control of soil erosion and sedimentation of downstream waterbeds, amelioration of climate, protection against weather damage, groundwater recharge, purification of air and water by acting as a "sink" for greenhouse gases (including carbon dioxide in logged forests if the timber extracted from them is not burned), conservation of genetic resources and biological diversity, and generation of recreational benefits and aesthetic values. While not all tropical forests provide these services to the same degree or even exclusively, ignoring them results in a lower return on forest investment and causes environmental problems that can combine to make timber production a socially undesirable and ecologically unsustainable industry.

Examples, positive and negative, abound. The 1982 forest fire in eastern Borneo resulted in enormous losses of timber and non-timber production (estimated at over $6 billion—more than the export value of all forest products from Indonesia over two years) as well as increased soil erosion, local climatic changes, and extinction of species (Leighton and Wirawan 1986). Its severity, it is believed, resulted from the extensive logging carried out in the area. A study of the Tai forest of the Ivory Coast, which has the world's highest deforestation rate (7 percent per annum), found that rivers flowing from primary forests release twice as much water halfway through the dry season, and between three and five times as much at the end of the dry season, as do rivers from a coffee plantation zone. Had the watershed function of the forest been evaluated at the social scarcity value of water, less deforestation, including forest conversion to coffee plantations, might have been allowed. On a positive note, benefits of $30 million were estimated from a $1.8 million investment in reforestation of the watershed of Poza Honda reservoir in Ecuador, and $3 million to $10 million per annum in additional economic activity, including multiplier effects, were generated from the expenditures of institutions doing research in the tropical forests of Costa Rica.

The order of magnitude of these estimates indicates the potential benefits from taking environmental services into account in forest investment and management. Clearly, not all environmental services are compatible with logging. Conservation of soil and water, on the one hand, can be compatible with logging provided that adequate ground cover is maintained at all times and logging methods are regulated (e.g., clear-cutting of steep slopes is prohibited). Full conservation of genetic resources, on the other hand, is generally incompatible with logging, although it is fully compatible with conservation of water and soil resources, preservation of wilderness, and nature-oriented tourism.

An additional service that biomass-rich forests provide is the sequestration of carbon, which, oxidized in the greenhouse gas carbon dioxide, may lead to global warming. Tropical moist forests embody 55 percent of the world's organic carbon in living biomass, equal to 188 billion tons, which is 34 times the carbon released from fossil fuel consumption annually. Carbon sequestration, though a global benefit not captured by the countries in which tropical forests are found, has a potentially high value in the context of a possible international climate convention that would limit the allowable carbon emissions and create opportunities for trading permits for carbon emissions and sinks.

MULTIPLE-USE MANAGEMENT

When managed at all, tropical forests tend to be managed for a single use, usually timber, although in recent years there has been a modest expansion of national parks and nature reserves managed for conservation and recreation, and village woodlots maintained for fodder and fuelwood. This study's fundamental premise is that multiple-use management of tropical forests holds the key to economic profitability, social acceptability, and ecological sustainability. This form of management and investment will, in turn, ensure the long-term sustainability of tropical hardwood supplies.

Multiple-use management recognizes and attempts to evaluate all possible uses of tropical forests, without assuming that all uses should occur everywhere. Multiple-use management usually involves a full evaluation of all forest goods and services, and posits a set of criteria for selection of the optimal use or combination of uses to be permitted in a given forest area. Management systems are needed because multiple use of the same forest area may involve multiple users, numerous and conflicting management objectives, multiple time frames, and negative interactions among uses and users. Multiple-use management evaluates the complex ecological, economic, and social interactions and trade-offs among uses to determine the optimal use mix based on the criterion of maximization of net present value to the owner or decision maker.

One version of multiple-use management that simplifies the choice among large numbers of uses and their combinations is dominant-use management. This approach selects the use with the highest net present value as primary for a given forest area, and adds secondary uses to the extent that additional benefits exceed the extra costs. The benefit-cost calculations include the positive and negative interactions among these uses. Thus, certain areas are designated as timber forest in which collection of non-timber goods is allowed, while other areas are managed for watershed protection with limited logging and collection of non-timber goods. The negative interactions between logging and genetic resource conservation require that they be spatially separated, unless special extraction methods are used that may be unprofitable except for extremely valuable wood. For other uses, the conflicts are more apparent than real. For example, conservation of soil and water, which imposes certain restrictions on timber harvesting, is critical to sustainable timber management over the long run.

Seeking to maximize the net present value of tropical forests presumes the ability to estimate and compare the benefits and costs of all forest goods and services, in addition to quantitative knowledge of their ecological interactions and trade-offs. The lack of adequate and reliable information on non-timber goods and services makes such estimation difficult, though not impossible. Moreover, many of these goods and services are not traded, and therefore not valued, in formal markets. Rather, they are either consumed locally (non-timber goods) or are generated as intangibles or side effects outside the domain of markets and external to forest management (e.g., watershed protection and genetic conservation). Even timber, a major internationally traded commodity, is grossly undervalued, and its price is often subject to policy distortions such as taxes, subsidies, and tariffs that bear no relationship to social or environmental benefits and costs.

On the positive side, sophisticated methods have been developed in recent years for evaluating goods and services for which market prices either do not exist (most non-timber goods and all services) or do not reflect their true social scarcity value (some non-timber goods and quality hardwood timber). Selected valuation methods and applications reviewed in this study include productivity changes, cost-effectiveness, replacement and compensation cost, property values, and travel cost approaches.

SILVICULTURE AND LOGGING

Multiple-use management also requires certain modifications of existing silvicultural practices and logging technology, if it is to be implemented at the stand level. While it is difficult to develop separate management strategies for timber and non-timber species within the same forest, silvicultural methods should be modified to reduce undue damage to desirable non-timber species if they do not compete with timber species for space and light. The reverse should be practiced when non-timber harvest is the dominant use of the forest for economic or social reasons.

More critically, logging technologies must be carefully selected and controlled to reduce damage to both timber and non-timber species and ensure their regeneration. Clear-cutting of large areas is certainly detrimental to both non-timber goods and services and to hardwood forest regeneration. Selective logging stimulates growth of young regeneration by improving light conditions and nutrient availability, but also damages neighboring tree crowns and understory vegetation, thereby reducing the photosynthetic potential of the residual stand and the survival of seedlings. Logged-over areas often suffer from erosion and soil compaction by heavy machinery. Such damage can be lessened through carefully planned and well-executed logging. In addition to decreasing damage, these measures can lower overall logging costs. For example, cutting vines and climbers before logging can reduce damage to the stand, while leaving buffer strips of forest on both sides of streams can restrict downstream damage significantly. Finally, logged-over areas that fail to regenerate adequately and have low opportunity costs, especially on the sides of logging roads, could be silviculturally managed through enrichment planting to reduce runoff and erosion as well as to enhance the timber yield of the ensuing crop. These silvicultural methods can also produce fuelwood and fodder for local populations, thereby minimizing social pressure on natural forests.

PLANTATION FORESTRY

Forest plantations are widely regarded as economically more attractive investments than sustained yield management of indigenous forests. However, plantation forestry as currently practiced cannot compensate for the timber shortage that results from natural forest degradation, let alone the loss of non-timber goods and services. Of the estimated 12 million hectares of forest plantations in the tropics, almost 80 percent consists of fast-growing species used for fuelwood, paper pulp, wood chips, and low-quality timber; only 2.5 million hectares have been planted with slower-growing hardwood species for the production of high-quality sawnwood. The relative success of monoculture plantations in the seasonal tropics may not be transferable to moist forests, where the lack of a dry season can accentuate pest and disease problems in single-species stands. The clear-cutting method of harvesting used in these plantations exposes the soil to erosive rainfall and winds, which cause substantial soil disturbance on fragile lands. In some countries, the establishment of industrial forest plantations serves primarily as a legal mechanism to excise blocks from protected forest reserves. Thus, plantations may well accelerate deforestation rates, rather than provide an alternative source of timber.

In light of the problems of single-species plantations and their insignificant contribution to hardwood timber supplies from the humid tropics, the potential of mixed-species plantations should be explored. Many timber species, such as many dipterocarps of Asia or the many mahogany species of Africa, will not establish or grow well in the early stages without shade cover or nurse crops. Mixed-species plantations, by incorporating more than one type of ecological guild, can simulate the successional processes of the natural forest. Density-dependent mortality in natural forests and pest outbreaks in single-species plantations suggest that mixed-species plantations might be less vulnerable than monocultures to pests and disease. Mixed-species plantations also have the potential for incorporating non-timber species, such as rattan and fruit trees, without detriment to timber production when predictability, sustainability, and social pressures are taken into account. On the negative side, mixed-species plantations require more complex silviculture, higher harvesting costs, and longer rotations, but they are also likely to produce more valuable goods and services than single-species plantations. They would generally be more labor-intensive and hence more suitable to factor endowments in developing tropical countries. There is a need for experiments to test the technical and economic feasibility of mixed-species, multi-purpose forest plantations in the humid tropics.

CONSERVATION OF GENETIC DIVERSITY

Mixed-species plantations, whatever their potential for producing timber, non-timber goods, and environmental services, can neither conserve genetic resources nor preserve natural wilderness. In this respect the natural rainforest is irreplaceable. Any reduction in natural forests inevitably leads to some extinction and attrition of genetic diversity. The aim of conservation must therefore be to optimize species diversity, rather than to preserve everything, which in practice is impossible. Specialized animals and plants can usually survive in relatively small areas when they are maintained in a completely unexploited, unmodified state. Generalized species often require larger areas, but cyclical, selective exploitation is often not harmful to them.

A carefully planned conservation strategy could preserve the great majority of both specialized and generalized species. The minimum individual conservation area for each habitat should be at least 5,000 hectares, connected by corridors of managed natural production forest. Together, they should comprise a total conservation area of at least 100,000 hectares. They would, of course, have to be environmentally heterogeneous to ensure adequate representation of genetic resources. In principle, a few large environmentally heterogeneous preserves are preferable to many small environmentally uniform preserves, and have the added advantage of preserving wilderness and aesthetic values. Selectively logged corridors connecting protected reserves must be managed so that keystone food plants, such as figs, are preserved. Finally, the full range of sites in each climatic zone must have adequate representation of the zone's genetic resources, with first priority given to areas of high species endemism and species richness. Plant species diversity is greatest in Latin America with 90,000 species, followed by tropical Asia with 45,000, and Africa with 35,000. Richest diversity is found in the foothills of the equatorial and northern Andes, followed by the forests of Borneo, the forests of Cameroon, the foothills of southern and western New Guinea, and the lowland forests of peninsular Malaysia.

MULTIPLE-USE MANAGEMENT:
CONSTRAINTS AND OPPORTUNITIES

Extensive review of the literature, consultation with technical experts in the field, and preliminary assessment of the feasibility of incorporating non-timber goods and services into natural forest management and mixed species plantations all suggest considerable potential for multiple-use management of tropical forests. However, the extent to which this potential can be realized depends on four factors: (1) institutional reforms, (2) government policies, (3) international cooperation, and (4) further research and development necessary to fill the gaps in knowledge.

Institutional Reform

Perhaps the greatest constraint to efficient multiple-use management of tropical forests and optimal forest investment is the general uncertainty and insecurity of ownership of forest resources. Historically, most tropical forests were communal or tribal property to which the members of the community or tribe had customary rights of access and use. During the past 150 years, over 80 percent of the forests of tropical Africa, Asia/Pacific, and Latin America have been brought under government ownership. Special legal status has been given to particular areas such as forest reserves, protected forests, and national parks.

Governments have generally been unable to enforce their ownership of the forest. As a result, tropical forests, with the exception of a few national parks, have reverted to quasi-open access with pervasive encroachment, squatting, log poaching, slash-and-burn agriculture, and unregulated conversion to other uses. At the same time, governments continue to award concessions to logging companies at truly concessionary terms, often on political grounds, and fail to enforce harvesting and replanting regulations. Logging companies operating under the constant threat of encroachment and interference by communities deprived of their customary rights to forest products tend to adopt destructive strategies that result in mining, rather than managing, their forest concessions. This results in a pervasive climate of lawlessness, uncertainty, and insecurity of tenure for all parties concerned—governments, logging companies, squatters, and local communities. No single party has sufficient control and incentive to conserve the resource base or invest in its management and enhancement.

The solution to the problem lies in establishing well-defined, exclusive, and enforceable rights over forest resources and providing rural and forest-dwelling populations with better alternatives for earning a living. Forestlands with no significant externalities (spillover effects) can be safely distributed and securely titled to individuals. Forestlands with localized externalities, such as local watersheds and extractive reserves, can be made communal property provided that a small, cohesive community able to effectively manage them can be identified. Forestlands with national or international externalities, such as major watersheds and nature reserves, should remain under state ownership. Secure ownership and investment incentives commensurate with environmental services are needed. Forest investments and eventually timber supplies from the private sector would then expand to more than compensate for any reductions in public investments.

Government Policies

Despite the proposed transfer of part of government-owned forests to individuals and communities, this study envisions a somewhat expanded role for governments in the forestry sector: (1) as owners of a reduced but still substantial portion of the country's forests, which could be more intensively managed; (2) as regulators of economic activity; and (3) as development agents.

As forest owners, governments can increase the efficiency of resource use and their own share of revenues by awarding concession leases through competitive bidding. Duration and scope of concessions can be increased to internalize non-timber forest products and services and forest regeneration for subsequent felling cycles. Alternatively, governments may assume the full responsibility of forest management and grant logging rights to accredited logging firms through public auctions. Furthermore, governments' decisions to invest in protection, management, and enhancement of state-owned productive and protective forests should be based on strict criteria of social profitability.

As regulators of economic activity, governments may set standards for forest use by the private sector and local communities to promote wider objectives such as equity, stability, and national security; to mitigate market failures; and to provide public goods. In this respect, governments must ensure that taxation does not promote overharvesting and waste (e.g., incentives for forest conversion to ranching).

More positively, governments could make long-term financing available for forest investments, to moderate the effects of myopic capital markets while providing forest investors with investment incentives commensurate with the environmental and social services they provide and for which they receive no return through the market. Finally, governments must ensure that macroeconomic and sectoral development policies do not indirectly introduce unintended incentives for overexploitation and disincentives for forest investments (e.g. excessive protection of local timber processing, crop and livestock subsidies, or overvalued exchange rates).

International Cooperation

While institutional and policy reforms by individual governments would go a long way toward rationalizing the exploitation of forest resources and improving the economic environment of forest investments, international involvement is important. Without international cooperation, their efforts may be frustrated by the continued supply of undervalued timber from other countries that continue to mine, rather than manage, their natural forests. All tropical timber exporting and importing countries stand to benefit from the assured stability and sustainability of the timber trade that will result from full valuation of timber and non-timber goods and environmental services. They will also benefit indirectly from the conservation of genetic resources and improved economic and natural environments. These benefits would be larger and more likely to materialize if there is coordinated action by producing and consuming nations to rationalize the use of tropical forests. Potentially significant, in this respect, are international conventions for tropical forests and biodiversity currently under discussion in the context of the United Nations Conference on Environment and Development (UNCED).

Research and Development

Our knowledge and understanding of how the tropical forests function is severely limited. For instance, with the exception of limited and fragmentary information, we do not know how the different rainforest plant and animal species interact with each other and how their interaction responds to natural and human disturbance. We do not yet have good estimates of the economic value of non-timber products and environmental services provided by representative tropical forest types and regions. Without information on such basic interaction parameters and values, it is difficult to select the dominant use or the combination of uses that would yield the maximum social value for a particular tract of tropical forest. Clearly, much more long-term natural and social science research is needed before we can develop models of sustainable tropical forest management. However, we do know enough to determine that current policies and practices are promoting destructive and unsustainable uses of tropical forests and to recommend major policy reforms. Basic policy reform is essential to the reinstatement of the tropical forest as a valuable economic, ecological, and social asset.

ORGANIZATION OF THE STUDY

The study is organized into fourteen chapters. Following this introduction and overview, chapter 2 describes the current situation of the tropical forests in Latin America, Africa, and Asia/Pacific, including the regional rates and causes of deforestation. The chapter concludes with a discussion of existing and future trends of trade and consumption of tropical hardwood timber. Chapter 3 reviews current attempts to manage the natural forests in tropical areas for sustained harvests of timber, while chapter 4 documents the fact that forests, despite serving as sources of valuable timber, have been grossly undervalued by governments throughout the tropics. Chapter 5 describes products other than high-quality hardwood timber provided by tropical forests, including non-timber wood and non-wood products. These products too are grossly undervalued, and this chapter attempts to delineate the reasons for their neglect by governments and forest users, as well as provide some estimates of the monetary and economic worth of these goods and services. Chapter 6, in similar fashion, examines the situation of environmental services provided by the tropical forests, the benefits of which must be taken into account in considering the value of the resource by governments in the countries producing hardwood timber.

Chapter 7 develops a framework for the economic analysis of multiple-use forest management. The chapter also examines some of the technical problems of valuation in dealing with products and services without well-defined markets to set prices. Chapter 8 is a brief review and analysis of silviculture and logging technology for multiple-use management, including enrichment planting. Chapter 9 explores the silviculture and economics of plantation forestry, including single- and mixed-species plantations.

The following four chapters look at a series of broader issues related to the problems analyzed in previous chapters. Chapter 10 looks into the conservation of genetic resources. Chapters 11, 12, and 13 identify, respectively, institutional reforms, government policy changes, and international cooperation needed to implement tropical forest management for multiple use. Chapter 14 summarizes the identified gaps in knowledge and provides recommendations for further research and development.

CHAPTER 2

Tropical Forest
Resources and the
Timber Trade

EVERY DAY, forested regions throughout the tropics of Latin America, Africa, and Asia lose hundreds of hectares of trees as a result of deforestation. Numerous studies conducted over the past decade indicate that the world's tropical forests are being depleted at an unsustainable rate. If present deforestation rates continue, timber reserves in many countries could drop precipitously. Countries such as the Ivory Coast and Nigeria, which have extremely high deforestation rates and only moderate amounts of forest, could actually lose *all* of their forests by 2025 (WRI 1988). The ongoing depletion of the world's tropical timber reserves also signals serious danger for the survival of the total timber trade, which was worth approximately $50 billion in 1985 (WRI 1988).

World demand for tropical timber has grown significantly in the last several decades. While many factors influencing timber demand can be identified, population growth and increased economic development play a leading role. Population pressures have forced more people to rely on marginal forest land that would be better left undisturbed. Also, as economic growth spreads throughout the developing world, demand for timber for construction and other purposes rises. The total annual volume of tropical hardwood timber removals increased by 73 percent from 78 million cubic meters to 135 million cubic meters between 1965 and 1980, with removals projected to increase to 236 million cubic meters by the year 2000 (see tables 2.1 and 2.2). Tropical hardwood imports increased by 14 times in the 30 years before 1980 (see table 2.3). Although cyclical economic downturns may

16

Table 2.1
TRENDS IN TROPICAL HARDWOOD REMOVALS, 1965–1987
(IN MILLIONS OF CUBIC METERS)

	1965	%	1975	%	1985	%	1987	%
All Tropics	77.746	100	113.747	100	134.418	100	143.836	100
Africa	11.666	15	14.296	13	17.379	13	16.695	12
Asia/Pacific	51.348	66	77.647	69	85.955	64	94.714	66
Latin America	14.732	19	21.804	19	31.084	23	32.427	23

BASED ON: Grainger (in press).

Table 2.2
PROJECTED NET REMOVALS OF HARDWOOD LOGS IN TROPICAL COUNTRIES
FOR THE YEAR 2000
(IN MILLIONS OF CUBIC METERS)

	(A) Operable Forests	(B) Plantations		(A + B) Total Net Removals
		Low Yield	High Yield	
Africa	23	1	7	31
America	52	3	22	77
Asia & Oceania	111	1	16	128
Total Net Removals	186	5	45	236

BASED ON: Erfuth 1984.

Table 2.3
MAJOR TROPICAL HARDWOOD IMPORTS TRENDS, 1950–1987
(IN MILLIONS OF CUBIC METERS, ROUNDWOOD EQUIVALENT VOLUME)

	1950	%	1960	%	1970	%	1980	%	1987	%
Total Imports	4.364		15.759		48.832		61.226		62.822	
Japan	0.209	5	4.744	30	20.652	42	19.300	32	19.208	31
EEC	1.663	38	5.294	34	8.549	18	10.277	17	8.171	13
USA	0.680	16	1.989	13	3.144	6	2.681	4	3.461	6
Gulf Nations	na	—	na	—	na	—	0.373	1	0.728	1
Other Asia	na	—	na	—	6.658	13	13.571	22	12.181	19

BASED ON: Grainger (in press).

cause temporary decreases in international trade, as appears to have happened in recent years, total long-term demand is likely to grow rapidly as a consequence of population growth and rising incomes, particularly in developing countries. However, if trends continue—current logging methods, deforestation rates, underinvestment, mounting opposition by indigenous populations and environmental groups—timber prices will rise at a rate governed by the elasticity of substitution between tropical and temperate varieties. Unsustainable management practices characterized by shortsighted planning, the undervaluation of forest goods and services, insecure ownership of forest land, and inefficient logging concession and taxation systems have resulted in a dangerous depletion of the tropical forestry resource base. In this case, rising timber prices do not simply reflect the increasing scarcity of the resource but, rather, its wasteful demise.

Tropical timber supply is affected by four interrelated, fundamental factors: (1) the condition of the resource base—the tropical natural forests; (2) current and expected prices associated with the tropical timber trade; (3) levels of investment in natural forest regeneration and in forest plantations; and (4) attitudes, perceptions, and reactions of local populations and environmental groups toward logging and other forest activities. Several factors indicate that tighter supplies and increasing prices of tropical hardwoods will be the likely future consequences of current patterns of timber production and rates of forest investments. Trade patterns also point to tighter supplies and higher prices. In general, log prices are forecast to increase in real terms reflecting increasing supply constraints along with a boost in demand associated with higher rates of economic growth, particularly in Asian countries (World Bank 1989). While many of the indigenous tropical hardwoods that are exported are also consumed in national markets, the higher grades are primarily exported. Domestic markets, however, may be supplemented by exotic species (e.g., *Gmelina, Araucaria*) grown in plantations. In the future, it is likely that significant amounts of this wood will be used for local construction materials, especially in view of the shift of timber-processing industries to log-producing countries (World Bank 1989).

THE CONDITION OF THE RESOURCE BASE

The world's forests and woodlands probably covered 5 billion hectares at one time (WRI 1990). By the mid-twentieth century, they had declined 20 percent to approximately 4 billion hectares due to the increasing demand for agricultural land, pastures, and settlements for rapidly increasing populations. Until that time, the greatest changes in vegetation cover occurred in the temperate regions. Over the past three decades, however, deforestation in the tropics has been far greater than in temperate regions, where a small net increase in forest area has occurred. Slightly more than one-half the world's forests are in developing countries. Worldwide, forest covers approximately one-third of the land area (see table 2.4). In developing countries, deforestation has surpassed reforestation rates by 10 to 20 times in recent years, while temperate forest areas in Europe, Asia, and Oceania have grown somewhat, and are only slightly decreasing in North America (WRI 1987, 58–59). As owners of more than 80 percent of the world's tropical forests, developing-country governments have been unwilling or unable to capture more than a small fraction (10 to 50 percent) of the stumpage value or scarcity rent of the resource (Repetto and Gillis 1988; Vincent 1990). The undervaluation of tropical hardwood timber and its resource base, combined with the overvaluation of the net benefits from forest conversion to other uses (e.g., farming, ranching) has led to excessive deforestation and failure to implement natural forest management.

Deforestation is defined in this report as conversion of natural forest areas to other uses including (artificial) forest plantations, agriculture, and wasteland (following Lanly [1982]). Thus, the deforestation statistics given below include (1) conversion of forest to non-forest cover, such as grassland, non-tree agriculture, wasteland, and secondary

Table 2.4
FOREST AREAS IN SELECTED COUNTRIES
(IN THOUSANDS OF HECTARES)

	Forest Area	% Total Land
Countries with Large Areas of Forest		
Brazil	514,480	60.5
Zaire	177,590	75.7
Indonesia	116,895	61.4
Countries with High Percentage of Forested Lands		
Peru	70,640	54.9
Bolivia	66,760	60.8
Central African Republic	35,890	57.6
Botswana	32,560	54.2
Papua New Guinea	38,175	82.7
Malaysia	20,996	63.7
Countries with Low Percentage of Forested Lands		
Mali	7,250	5.8
Kenya	2,360	4.0
Haiti	48	1.7
Uruguay	490	2.8
El Salvador	141	6.7

SOURCE: WRI 1990.

scrub occupying abandoned swidden sites; (2) conversion of natural forest to perennial tree-crop agriculture—rubber, oil palm, cacao, fruit trees, spices, coffee; and (3) conversion of natural forest to artificial forestry plantations for the production of timber, paper pulp, cellulose, fuelwood, and charcoal. It should be noted that categories (2) and (3) in effect constitute a conversion of complex, natural forest to simple, artificial forest, which under optimal conditions may nevertheless continue to provide in part some of the environmental services of natural forests, such as watershed regulation, climatic and atmospheric amelioration, and soil protection. However, with the possible exception of rubber and fruit tree plantations, category (3) is the only form of conversion that continues to provide significant amounts of wood products. Natural forest conversion to plantations constitutes a very small part of total deforestation.

Deforestation statistics compiled in 1981 by the Food and Agriculture Organization of the United Nations (FAO) and the United Nations Environment Programme (UNEP) present only a partial picture of the extent of both economic and ecological damage occurring in tropical forests, because they omit effects such as overgrazing, fire damage, overharvesting of timber and fuelwood, and selective encroachment, all of which lead to forest degradation. The immediate causes of tropical deforestation range from encroachment by migrant, landless farmers to conversion of forest to pastureland to the cutting down of trees for fuelwood by area villagers. Estimates of annual tropical deforestation in the range of 11.3 million hectares, based on the 1980 FAO data, have recently been increased dramatically as a result of new information on higher rates of forest loss in Brazil, Costa Rica, India, Myanmar, the Philippines, and Viet Nam; revised estimates are 79 percent higher, indicating that tropical deforestation could total 20.4 million hectares annually (see table 2.5).

Another major study of deforestation in the tropics (Myers 1980) estimated total destruction at 24.5 million hectares per year (not including fallows or open space). Myers's larger figure is mainly due to the different definition and set of criteria used. Because the study's emphasis was on the genetic rather than the environmental or timber values of the forest, a broader set of criteria was used, including many forms of degradation or disturbance, while the FAO report considered only outright destruction of the forest. For this study we have chosen to use FAO statistics (including available revisions) because they are generally regarded to be the most accurate current assessment and because they are more directly useful for examination of the timber issues dealt with in the study. Nevertheless, Myers's estimates give better representation of the value of non-timber goods, which may affect

Table 2.5
AVERAGE ANNUAL DEFORESTATION BY REGION, 1980s
(IN THOUSANDS OF HECTARES)

Region	Closed Forests		Open Forests		Total Deforestation	
	Area	%	Area	%	Area	%
Tropical America	10,909	1.11	1,313	0.34	12,222	0.90
Brazil[a]	8,000	2.23	1,050	0.67	9.050	1.76
Tropical Africa	1,359	0.62	2,406	0.52	3,765	0.55
Zaire	182	0.17	188	0.26	370	0.21
Tropical Asia	3,931	0.96	57	0.07	3,988	0.81
Indonesia[b]	900	0.79	20	0.67	920	0.79
TOTAL	16,199	1.00	3,776	0.41	19,975	0.79

SOURCE: WRI 1990.

a. Figures are for 1987, one year before tax incentives for forest settlement were scaled back.
b. Annual averages for 1979–1984.

the future sustainability of timber production forests through the negative impact that this loss has on the afforestation of forests by rural communities.

Regional Variations in Deforestation

There are important differences in rates, causes, and effects of deforestation from one country or region to another, due to the relationships between population density and growth rates, levels and rates of economic development, distribution of wealth and land, tenure systems, and cultural attitudes. Overall, however, about 45 percent of the deforestation of closed forests worldwide can be ascribed to unsustainable shifting cultivation by migrant farmers (Lanly 1982). There are at least 200 million traditional and immigrant shifting cultivators living within the tropical forests, and the number is growing rapidly (Myers 1984, 156).

Presently, deforestation in tropical Asia results primarily from encroachment by lowland populations into upland forests using unsustainable, degradative agricultural techniques, and partly from planned transmigration and resettlement. Especially large areas have been deforested in Myanmar, India, Indonesia, Nigeria, the Ivory Coast, and Zaire (see table 2.6).

In Africa, conversion and degradation have been particularly severe in semiarid West and East Africa, where supplies of fuelwood, poles, forage, and other non-timber wood products that rural households need have dwindled. Approximately 62 percent of the deforestation of the world's open tropical forest and woodlands has occurred in Africa. Shifting cultivation is the cause of more than 70 percent of deforestation of the closed forests in Africa (Lanly 1982).

As much as 55 percent of the total deforestation of closed forests in tropical Africa occurs in the nine countries of West Africa, with 45 percent of that occurring in the Ivory Coast and Nigeria. Over one-half of the world's forest loss occurred in the West African countries of the Ivory Coast, Nigeria, Liberia, Guinea, and Ghana alone. According to the FAO (1981a, b, c), the rate of forest loss in West Africa in 1980 was estimated to be seven times the world average because of the exceptionally high deforestation rates in the Ivory Coast (5.9 percent) and Liberia (2.2 percent). The Ivory Coast had the world's highest deforesta-

Table 2.6
DEFORESTATION IN TROPICAL COUNTRIES, 1981–1985
(IN THOUSANDS OF HECTARES)

	Closed Forest Area, 1980	Annual Rate of Deforestation	Annual Area Deforested
Higher than Average Deforestation Rates; Large Areas Deforested			
India	36,540	4.1[a]	1,500[a]
Myanmar	31,941	2.1	677
Nigeria	5,950	5.0	300
Ivory Coast	4,458	6.5	290
Relatively Low Deforestation Rates; Large Areas Deforested			
Zaire	105,975	0.2	182
Madagascar	10,300	1.5	150
High Deforestation Rates; Small Areas of Remaining Forest			
El Salvador	141	3.2	5
Jamaica	67	3.0	2
Haiti	48	3.8	2
Low or Moderate Deforestation Rates; Small Areas Affected			
Dominican Republic	629	0.6	4
Equatorial Guinea	1,295	0.2	3
Ethiopia	4,350	0.2	8

SOURCE: WRI 1990.

a. Annual deforestation for 1975–1982.

tion rate by far, estimated to average nearly 7 percent annually in the 1980s (Repetto 1987). Deforestation rates in the rest of West Africa have been moderate—Congo (0.1 percent), Gabon (0.1 percent), Cameroon (0.4 percent), and Ghana (0.9 percent)— but those rates are now increasing in Cameroon and Gabon. Deforestation rates in Zaire are usually assumed to be low, an assumption that may be incorrect: Reportedly, no reliable estimates exist for Zaire, which contains the largest tropical forest resource in Africa (P. Kio, personal communication). Deforestation in Zaire is almost entirely due to the expansion of shifting agriculture in response to population increases.

Deforestation in tropical Latin America has occurred because of population growth that leads to pressure to clear more land for farming, land speculation, and, quite distinct from Africa and Asia, the development of commercial ranches. The creation of pasture land for beef cattle is the major cause of deforestation in Central America (Leonard and Nations 1986) and in the Brazilian Amazon region (Schmink 1987). For the most part, ranching is made profitable only by massive government subsidization. Shifting agriculture is responsible for roughly 35 percent of total deforestation in tropical America (Lanly 1982). Nearly two-thirds of Central America's original forests have been cleared. There, up to three-fourths of the hardwood timber felled each year is burned rather than harvested. Commercial timber harvesting in Central America is important only in Honduras; but each year approximately $320 million worth of hardwood is burned or left to rot after cutting in that country (Leonard 1987). Deforestation is also driven by government policies that require land "improvement" (i.e., deforestation) to establish tenure rights.

In tropical America, Brazil accounts for 35 percent of the reduction in closed forest area, although deforestation rates are especially high (over 3 percent) in Paraguay, Costa Rica, and El Salvador and over 2 percent in Honduras, Bolivia, and Ecuador. Peru, Trinidad, and Tobago have the lowest deforestation rates, under one-half of a percent.

The U.S. Interagency Task Force on Tropical Forests concluded in 1980 that if present trends continued, the world's tropical forests outside of central Africa and the Amazon basin would be "nothing but scattered remnants" by the year 2025 (U.S. Interagency Task Force on Tropical Forests 1980). Assuming that the existing forest stock in developing countries is being removed at the rate of 15 to 20 million hectares per year, the World Bank Forestry Sector Policy Paper states, "At this rate, assuming no growth in demand, the remaining tropical forests will disappear in 60 to 80 years" (World Bank 1978). The World Wildlife Fund concludes that "a reasonable figure for annual loss of tropical forest in which all or most of the ecological value is lost, is 11–15 million hectares. Taking the lower figure this approximates close to 31,000 hectares a day or 22 hectares (more than 20 football fields) a minute" (WWF 1987, 22).

The FAO report notes three elements that could reverse these tendencies in the long term:

1. accelerated rate of urbanization in many tropical countries, which results in decidedly lower growth rates in the agricultural population as compared to the total population [for an example, see Vincent and Hadi (1991)];
2. natural reforestation of zones abandoned by shepherds and farmers, the importance of which, it is true, is very slight at the moment as compared to that of the deforestation; so slight, in fact, that it has not been taken into consideration in the [FAO] study;
3. intensification of farming methods, the organization of rural areas, the reservation of a permanent area of productive or protective forests (national parks), are all actions that are either still at the embryonic stage or at a greatly reduced level but which will tend to develop progressively in a growing number of countries (Lanly 1982, 77).

The following additional elements have worked in developed countries. Their impact in developing countries would depend on the success of institutional reform and rural development:

1. rising timber prices, as the more easily accessible forests are used up, leading to decreased demand and/or higher investment in timber production;
2. income growth, which reduces the use of the forest as a source of fuelwood and fodder and increases the demand for environmental services of forests; income growth may lead also to increased demand for high quality timber, but this may stimulate more investment in natural forest management if tenure problems are resolved;
3. technological developments, such as improved plantation techniques and silviculture;
4. reduction of logging and sawmill wastage, which could increase yield by up to 40 percent;
5. improved wood utilization; and,
6. development of wood substitutes.

Extent of Logging

Although loggers usually limit their attention to a few species and extract only a limited number of individual trees per hectare, logging can cause considerable damage to the remaining forest and disrupt, reduce, or change certain environmental services that the natural forest provides. Once a concession is logged for its valuable commercial timbers, it usually requires at least 40 years before it can be logged for those species again, even if selectively logged. Tropical America has already logged over 10 percent of its productive, closed broad-leaved forest. The proportions in tropical Africa and tropical Asia are now more than 27 percent and 49 percent, respectively (Lanly 1982).

Although selective logging itself does not usually result in outright deforestation, the ecological and biological complexity of the forest may never recover if logging continues. The highly selective character of the timber trade has led to the severe depletion and may eventually lead to the extinction of some high-value species (mahogany is currently under consideration for inclusion under CITES, the Convention on the International Trade of Endangered Species of Wild Flora and Fauna [Vincent, personal communication]). Logging activities also create new road systems that facilitate the spontaneous colonization of hitherto inaccessible forest, with consequent large-scale clearing by the new migrants. As the FAO study by Lanly estimated,

> The existing "reserves" [used here in the mining sense of "reserves"] . . . correspond, therefore, to 178 years of logging at current levels and conditions in tropical America; 104 years in Africa and 42 years only in tropical Asia . . . assum(ing) that all the productive forests will become progressively economically accessible and that they will not be cleared before logging—two hypotheses that lead to an overestimation of these "reserves"—and that the VAC (volume actually commercialized) will not increase, which in general is false, since the regression of productive forests generally stimulates the logging of so-called "secondary" species—an hypothesis which leads to underestimation of these "reserves." (Lanly 1982, 61)

Deforestation results in the loss of considerable amounts of valuable timber trees before they can be logged. Grainger converted deforestation rates into equivalent volumes of commercial timber and estimated that between 1981 and 1985 some 20.3 to 63.7 million cubic meters were lost, according to his low and high scenarios, respectively:

The high scenario figure was equivalent to about half the total removals of tropical hardwood saw-logs and veneer logs in 1980. Three-fifths of the loss occurred in Asia/Pacific. By 2020 the drain should decline to between 6.1 and 36.3 million cubic meters according to the low and high scenarios respectively. (Grainger 1986, 131)

Consequences of Deforestation

From an economic standpoint, a certain amount of conversion or deforestation is an efficient and productive use of resources that can lead to sustained levels of benefits. A certain degree of conversion of forests to agriculture is inevitable, and if done on better soils using proven farming systems, such conversion would be a logical shift. Some countries have explicitly embarked on deforestation as preparation for economic "take-off," rapidly converting their forest resources into what they hope will be productive industrial and agricultural assets (Burns 1986, 12).

While some tropical forest clearance has led to viable agricultural and ranch holdings, much has resulted only in degraded soils, increased erosion and siltation, shrinking habitats for large numbers of plant and animal species, and a growing shortage of wood products. All too often, agriculture on ill-suited soils produces just a few harvests for farmers before declining yields and increasing weeds force them to move on.

Large volumes of potentially valuable wood are simply burned by the encroaching populations. Furthermore, deforestation of watershed areas without compensating ground cover can result in devastating environmental problems. Over 30 years, the forest area of the Himalayan watershed declined by approximately 40 percent, contributing to shortages of wood, fuel, and food in the uplands and to floods and siltation in the downstream areas (Myers 1984, 263). The cost of repairing flood damage below the Himalayan catchments in India has been, on average, $250 million per year, in addition to the loss of production and livelihood experienced by millions (Spears 1982).

Major watersheds around the world are suffering from serious devegetation and erosion, which disrupt the water cycle and deposit extremely high loads of soil sediments into streams, lakes, and rivers (Myers 1984). These loads affect agricultural development, hydroelectric power, urban consumption, and other contributions to economic development. By failing to adopt appropriate forest management practices now, tropical nations jeopardize many other investments, and greatly increase the costs of future economic development.

THE TROPICAL TIMBER TRADE

Both domestic consumption and international trade of tropical wood products have increased over the past 40 years (Laarman 1988). While between 3 and 18 percent of the total wood production of different regions is traded in international markets (WRI 1988), exports as a proportion of regional roundwood removals in tropical countries range from less than one-tenth in Latin America to more than one-half in Asia/Pacific, with Africa falling between the two. Between 1965 and 1980, this proportion increased substantially in Asia, but remained nearly the same in both Latin America and Africa, as shown in table 2.1. Tropical timber exports totaled about $8.7 billion in 1980 (Brazier 1982), generating a major portion of the foreign-exchange earnings of some developing countries. Logs and sawnwood are among the top three export commodities in nine out of twenty-six major tropical hardwood-producing nations. In 1980, the Central African Republic, Myanmar, and French Guiana received more than 20 percent of their foreign-exchange earnings from logs and sawnwood, while these products provided Cameroon, the Ivory Coast, Liberia, and Malaysia with 10 to 20 percent of their foreign currency (see table 2.7).

Developing countries as a whole are self-sufficient in solid wood forest products, although a $3.5 billion surplus in solid wood products in 1980 was almost exactly offset by imports of pulp and paper from developed nations (Grainger 1986, 16). Tropical moist forests accounted for only 13 percent of all world industrial roundwood production (logs, pulpwood, and poles), although another 6.6 percent comes from seasonally dry tropical forest. The bulk of roundwood exports from the tropics consists of one principal category of non-coniferous industrial roundwood, hardwood sawlogs, and veneer logs, most of which come from Asia.

Of all the roundwood felled throughout the world in 1980, approximately one-half was burned to provide heat and power, up from a figure of 40 percent in the late 1950s. In the near future, many more countries could experience increasingly severe deficits in fuelwood (WRI 1986, 67). Populations are considered to have fuelwood deficits when they are still able to meet their minimum fuelwood needs, but only by overcutting existing resources. An acute scarcity of fuelwood occurs when existing fuelwood resources have been depleted to the

Table 2.7
PRIMARY COMMODITY EXPORT PROFILES OF TROPICAL HARDWOOD-PRODUCING NATIONS, 1980
(IN MILLIONS OF U.S. DOLLARS)

	All Commodities	Logs & Sawnwood	All Commodities %	All Commodities Rank	Major Commodity	Export Value	% All Commodities
Africa							
Cameroon	1,321	144	11	4	Petroleum	405	31
Central African Republic	111	33	30	1	(2) Coffee	32	28
Ivory Coast	2,535	335	13	3	Cocoa	858	34
Liberia	597	73	12	3	Iron Ore	310	52
Asia/Pacific							
Myanmar	362	111	31	2	Rice	182	50
Indonesia	21,909	1,812	8	3	Petroleum	11,671	53
Malaysia	12,939	1,821	14	3	Petroleum	3,083	24
Latin America							
Bolivia	977	19	2	8	Non-ores	303	31
Brazil	20,132	145	0.7	—	Coffee	2,733	14
French Guiana	25	7	29	2	Fish	9	38
Other Asian Nations							
Singapore	19,375	449	3	7	Petroleum	4,809	25
South Korea	17,446	354	2	12	Clothing	2,855	16

BASED ON: Grainger 1986.

point where populations cannot obtain sufficient fuelwood even through overcutting (see table 2.8). In developing countries, the estimated proportion of wood felled and used for fuelwood varies from over 70 percent in Asia and South America to close to 90 percent in Africa (FAO 1985).

Nearly half of all trade in non-coniferous logs consists of exports from Asia/Pacific nations to Japan, which receives twice the volume that goes to other Asian nations for processing. Major flows in non-coniferous sawnwood occur among the Asian countries, with an equivalent volume exported from Asia to the European Community. The major flows of plywood are from Asia to Japan, the United States, and the European Community. The fourth major international tropical hardwood trade flow is in processed wood products (primarily sawnwood) from Latin America to the United States. Whereas prior to World War II world softwood supplies came from the same sources used now and in approximately the same proportions, hardwood supply has shifted dramatically. Before the war, more than half of sawn hardwood imported by Europe came from North America. After the war, West Africa became the major exporter, and now Southeast Asia is the leader (Gammie 1981, 38). It is predicted that within two decades Latin America will become the world market's major supplier of tropical wood (Grainger 1987a).

Before the 1940s, tropical hardwood imports were limited to high-quality decorative woods, such as mahogany from Central America, teak from Southeast Asia, and ebony from Africa. They were used mainly for furniture production. In subsequent years, international trade in tropical woods expanded significantly in response to demand for supplemental sources of utility woods in Europe, Japan, and North America. Many tropical timbers remain in high demand for their

Table 2.8
POPULATIONS INVOLVED IN FUELWOOD DEFICIT SITUATIONS
(IN MILLIONS OF INHABITANTS)

| | *1980* | | | *2000* |
	Acute Scarcity	Deficit	Prospective Deficit	Acute Scarcity or Deficit
Africa	55	146	112	535
Asia/Pacific	31	832	161	1,671
Latin America	26	201	50	512
TOTAL	112	1,179	323	2,718

BASED ON: Lanly 1982.

individual qualities, and have established markets in their own right. Imports by the United States have nevertheless declined in recent years, while Japanese and European imports have leveled off. This could be a temporary development due to a general slowdown in economic activity, and economic recovery could result in increased imports. Other markets could develop for tropical hardwoods, such as the People's Republic of China and the Persian Gulf countries. Currently, tropical hardwoods account for about one-third of E.C. (European Community) hardwood consumption.

Export trends indicate that while log exports have declined from developing countries, processed products have increased (see table 2.9). In 1980, Japan alone imported almost one-half and Europe nearly one-quarter of all tropical logs, sawnwood, and plywood/veneer exports; Japan's imports were predominantly logs. The United States is a relatively minor importer of tropical hardwoods, mainly in the form of sawnwood and plywood veneers (see table 2.10). Sawlogs and veneer logs continue to comprise two-thirds of total tropical hardwood exports (in roundwood equivalent). Latin America, Malaysia, and Indonesia have generated a significant increase in the last two decades in the proportion of processed timber exports. Indonesia, in particular, has rapidly developed plywood and sawnwood industries, becoming a major exporter of both commodities; Indonesian exports of plywood increased by 130 percent between 1983 and 1987 (World Bank 1989,

Table 2.9
TRENDS IN TROPICAL HARDWOOD EXPORTS, 1980–1987
(IN MILLIONS OF INHABITANTS)

	1980	1984	1987	% Change since 1980
Logs	38.475	26.282	28.690	− 25
Africa	6.547	4.911	3.594	− 45
Asia/Pacific	31.803	21.299	25.062	− 21
Latin America	0.125	0.072	0.034	− 73
Sawnwood	8.104	8.392	17.155	+112
Africa	0.717	0.617	1.426	+ 99
Asia/Pacific	6.578	7.208	14.691	+123
Latin America	0.809	0.567	1.038	+ 28
Plywood/Veneers	4.887	6.831	16.977	+247
Africa	0.284	0.373	0.515	+ 81
Asia/Pacific	4.231	6.153	15.745	+272
Latin America	0.372	0.305	0.717	+ 93

BASED ON: Grainger (in press).

Table 2.10
PERCENTAGE OF ALL IMPORTS BY PRODUCT AND SOURCE, 1980

	Japan	EEC	USA	Other Asia
Logs				
Africa	0	94	0	0
Asia/Pacific	100	6	0	100
Latin America	0	0	0	0
Sawnwood				
Africa	0	18	0	0
Asia/Pacific	100	76	33	0
Latin America	0	6	67	0
Plywood/Veneers				
Africa	0	15	0	0
Asia/Pacific	100	85	95	0
Latin America	0	0	5	0

BASED ON: Grainger, 1986.

344). Malaysia is the world's biggest exporter of hardwood sawnwood, tropical or temperate.

ROLE AND GROWTH OF PLANTATIONS

Tree plantations are regarded by many as the solution to the problem of decreasing tropical wood supplies and increasing deforestation. According to the World Bank, meeting world fuelwood demand in the year 2000 would require the planting of more than 50 million hectares of trees just for fuelwood, which represents a five-fold increase in the world's current rate of tree planting for all uses (World Bank 1980, cited in Allen and Barnes 1985).

The tropical countries involved in planting large areas are Brazil, Indonesia (which presently contains 72 percent of the high-grade tropical hardwood plantation area), India, and the Philippines. The ratio of the area of plantation to deforestation over the 1981–85 period was 1:10.5 in Tropical America, 1:29 in Tropical Africa, and 1:4.5 in Tropical Asia (Lanly 1982, 97).

Two-fifths of the plantation area of the humid tropics is dedicated to the production of fuelwood and other non-industrial purposes, and more than 40 percent of all industrial hardwood plantations are producing fast-growing, light hardwood species mainly for pulpwood and industrial fuelwood (see table 2.2). Less than one-fifth of all plantations in the humid tropics is dedicated to the production of the higher value hardwoods that are being extracted from the closed forests (Grainger 1986, 169). Thus, the commercially more valuable species are being depleted without replacement.

The FAO reports that more than 86 percent of plantations in 30 leading producer nations have been established since 1965. There has been a shift away from production of high-grade hardwoods, whose share of all plantings dropped from a peak of 54 percent in 1961–65 to only 13 percent in 1976–1980 (Grainger 1986, 170). Industrial plantations as a percentage of all plantings fell from 87 to 58 percent between the early 1960s and the late 1970s, due to increasing attention to the social and environmental aspects of forestry. Because of such changes, it is likely that planting of high-grade industrial hardwoods will continue at current levels, at best; Grainger concludes that tropical hardwood plantations will not be able to produce significant amounts of timber until the turn of the century, when removals will be relatively small compared to current removals from natural forests. In an optimistic scenario, his model forecasts that plantation production could compensate for the drain on tropical hardwood reserves due to deforestation by 2006, and in another less optimistic scenario, they would still only account for a third of the drain by 2021 (Grainger 1986, 174). (The role of plantations is discussed in greater detail in chapter 9.)

ATTITUDES OF LOCAL POPULATIONS AND ENVIRONMENTAL GROUPS TOWARD LOGGING

Local populations consider large-scale commercial logging a threat to their livelihood when the non-timber goods and environmental services on which they depend are devalued and degraded. As a result, community activist groups concerned with environmental issues have sprouted in many developing countries, including India, Brazil, Malaysia, and Kenya. Group members, often indigenous tribes or local villagers who subsist on traditional occupations such as rubber tapping, fishing, or nut gathering, consider that they are fighting for the economic and social survival of those they represent.

In India, tribal people have burned down teak plantations established by the government in the place of natural forests. The Chipko Andolan or "tree-hugging movement" in India began in March 1973 when residents of the village of Gopeshwar clung fearlessly to the trees that a timber company was preparing to fell. The movement has since spread to other parts of India, fighting to prevent government-licensed deforestation and logging (Caufield 1985, 156–57).

The rubber-tappers of the Brazilian Amazon began to rally against clear-cutting of the forest in the 1970s. Inspired by nonviolent methods of resistance, entire families have lain in front of bulldozers poised to trample the forest. Brazilian organizer Chico Mendes was assassinated in December 1988 following his relentless campaign to win the rubber tappers protection against eviction from the forests. Mendes was instrumental in establishing four land reserves for the tappers in the Amazonian state of Acre, as well as bringing international attention to the plight of the traditional inhabitants of the rain forest.

In Malaysia, opposition to loggers by local groups resulted in the erection of twenty-five roadblocks in Sarawak. Organized by the Penan and Kayan communities, the roadblocks were designed to thwart all removal of timber from the region's forests. The repeated arrests of Penan and Kayan representatives by the Malaysian government and the imposition of a new law in 1989 banning interference with logging operations have created an ongoing stalemate in that region (WRI 1990, 110).

Finally, the Greenbelt Movement, established in 1977 by the National Council of Women in Kenya, led to the planting of more than one million trees in 1,000 greenbelts throughout the Kenyan countryside. An additional 20,000 mini-greenbelts and 670 community tree nurseries have also been set up by local women's groups (Durning 1989). As part of the incentive system, the Movement pays women tree-tenders for each seedling that survives, leading to an 80 percent transplant survival rate (WRI 1990, 110).

SCENARIOS OF FUTURE TIMBER SUPPLY AND DEMAND

Present trends indicate increasingly severe pressure upon tropical forest resources from logging, cattle ranching, fuelwood collection, shifting cultivation, transmigration, and agricultural development. General developmental factors, most of which are difficult to foresee, assume greater significance when projecting timber production beyond a few years. These factors include demographic trends, political stability, exchange rates and foreign debt, national and international financial policies, cultural and sociological changes, agricultural productivity, and the emergence of competing materials. Because forests are also used for purposes other than timber production, other factors affecting deforestation and forest degradation rates, such as landlessness, lack of alternative employment opportunities, deficient land use planning, and policy distortions in favor of urban areas, must also be considered. Logging-access roads into forest areas frequently lead to spontaneous settlement, and consequent forest conversion, thus further limiting future timber harvests. Finally, the forest sector is susceptible to delays before management decisions are implemented. Therefore, when depletion of forest resources becomes serious enough to induce the establishment of plantations and increased forest protection and management, it is often too late to avoid a shortfall in removals for a considerable period.

A few general statements can be made regarding future timber supply and demand. Even if per capita consumption of forest products remained the same in the developing countries, total domestic consumption would double in 25 to 35 years given present population growth rates of 2 to 3 percent in most tropical countries. Rising living standards and higher domestic consumption would shorten this period. Traditional timber species are in increasingly short supply and will be replaced by lesser-known species and plantation-grown material, with smaller trees that have less heartwood volume (Zobel 1984, cited in Kauman 1987). As lowland forests become depleted of their higher-value commercial species, loggers will face higher costs as they move into more remote, less accessible regions. The social and environmental consequences of logging hillside forests will also escalate as these areas are prone to high erosion, thus affecting downstream agriculture and increasing the likelihood and costs of flood damage.

At present rates of consumption, the estimated fuelwood deficit will double by the turn of the century. While industrial roundwood production nearly tripled over the past 35 years, fuelwood production increased even more. Twenty countries in Africa, twelve in Asia, and seven in Latin America use over 80 percent or more of their total wood production for fuelwood, including both firewood and charcoal. About 60 percent of the people who are dependent upon fuelwood for cooking and heating (nearly 1.5 billion people) are cutting it back faster than it can be produced (WRI 1986, 66–68).

Grainger (1986) has developed a systems model of national land use containing mechanisms of deforestation, types of forest exploitation, and forest resources. In a 1987 publication, he concluded:

> Future trends in deforestation will most probably be determined by the way in which processes in the agricultural sector move towards equilibrium. This has major implications for strategies intended to bring deforestation under control since it indicates that the focus of action should be in the agricultural sector, rather than in the forest sector as in previous strategies, e.g., those of FAO (1985) and World Resources Institute (1986). (Grainger 1987b)

Grainger determined that deforestation could be greatly reduced given an increase in per hectare agricultural yield that remained only 0.5 percent ahead of per capita consumption.

The general pattern of future production of tropical hardwood timbers, as forecast by Grainger's "base scenario," indicates that Asia's tropical hardwood production and exports will peak in the 1990s, at which time Latin America will take over as the major world source of tropical hardwoods, supplying Europe and Japan, as well as North America. Latin America's exports and removals will start rising in the early 1990s and will peak before 2010. Exports then will start to decline as an increasing share of removals is consumed domestically, although even in 2020 the domestic market would be taking only a third of all removals compared with 89 percent in the late 1980s, prior to the rise in exports (Grainger 1987a).

> By the turn of the century, Latin America could be supplying almost two-thirds of all exports, or 77 million cubic meters, compared with a very small volume today. That means that removals will need to increase ninefold to around 96 million cubic meters over the next 20 years. Thus, the same kind of tidal wave of logging could sweep through the region as occurred in Southeast Asia. This could also last for about 20 years before supply constraints begin to be felt. . . . If the region's forests are exploited without much in the way of controls, and this seems likely on the basis of past experience, by 2020 the volume of commercial tropical hardwood reserves (in tropical America) would be less than a quarter of what they are today, and forests in most countries of the region except Brazil would show signs of acute depletion by about the turn of the century. (Grainger 1987a)

Grainger recognizes that while deforestation could be reduced to negligible levels within 40 years, a situation could develop whereby agricultural productivity does not increase as rapidly as predicted and encroaching cultivation becomes more and more widespread, resulting in large-scale forest clearance. In this case deforestation rates would become much higher than projected.

A 1981 study entitled "World Timber to the Year 2000" by the Economist Intelligence Unit (Gammie 1981) indicated that pre-1973 projections of future wood consumption were inflated, due in part to the use of economic boom years as baselines, and that, although economical extraction of wood resources would not continue forever, "tropical countries should be able to supply the traditional species currently in demand for some time to come." Most of the materials that compete with wood and wood products—concrete, plastic, bricks, steel, etc.—are energy intensive, whereas wood products require a great deal less energy. Therefore, while wood prices may increase at a higher rate than the general price index, it is doubtful that they will increase faster than their substitutes. According to this report's projections, the total exploitable tropical forest area will fall 16 percent, and annual offtake will increase threefold by the year 2000. Deriving information from World Bank data, the study estimates that total world wood supply will lag behind demand by about 8 percent in the year 2000, and by about 31 percent in 2025. That, of course, would cause prices to rise. However, the report notes that improvements in the recovery of wood fiber (particularly in developing countries), less wasteful logging practices, reduction in demand for paper due to microchip technology, increased utilization of more forest species (especially for production of reconstituted wood products), legislation, and greater reforestation can help meet some of the demand requirements (Gammie 1981).

Kauman (1987) concluded:

[S]till abundant resources of natural, tropical forests will be able to satisfy the likely demand for industrial wood for at least the next 25 years . . . if—and it is a big "if"—the catastrophic depletion by the collection of fuelwood and by deforestation to accommodate agricultural settlement, cattle ranching, and urban sprawl can be brought under control.

Kauman predicted that exports would increasingly yield to domestic consumption, and that higher incomes could raise local demand to the point where exports of industrial wood constitute a minority share of production.

Norman Myers (1984) took a dimmer view. He predicted that lowland forests in the Philippines, peninsular Malaysia, West Africa, and Central America would all be logged by the early 1990s. By 2000, he wrote in 1984, most of Indonesia's forests would be exploited, and a large portion of the Amazon would be claimed for cattle ranching and agriculture.

By contrast, Central Africa is sparsely inhabited and possesses abundant minerals. This reduces the incentive for governments to liquidate their forest capital in order to supply funding for various forms of economic development. Hence there could well remain large expanses of little-disturbed forest in Central Africa at the turn of the century. Similarly, the western portion of Brazil's Amazonia, because of its remoteness and constantly wet climate, might undergo only moderate change. However, as mentioned earlier, in Zaire deforestation may be occurring at a higher rate than is suggested by current, incomplete statistics.

More recent and more rigorous models of future trends in the timber trade predict a more gradual increase in global consumption of round-wood, due primarily to the increasing efficacy and role of plantations and natural forests in temperate regions (Vincent 1991). The FAO's predictions (1988)—the earliest of this recent group of studies and the most consistent with earlier projections of dwindling supply—indicate that consumption of industrial roundwood will increase at an average rate of 2.3 percent/year between 1990 and 2000, doubling in 30 years. The FAO projections concur with Kauman to the extent that consumption in developing countries increases faster than consumption in developed countries: Growth averages 3.4 to 3.5 percent/year in the developing economies, with the highest rates in South America at 4.5 percent/year, against 2.0 to 2.1 percent/year in the developed market economies. The picture painted by the FAO for processed products indicates highest growth of consumption again among developing market economies, but predicts that increases in production of processed products—sawnwood, wood-based panels, pulp, and paper—would match, or fall short of, consumption increases, indicating that the developing countries will increasingly import processed wood products.

The FAO projections do not include statistics on production of roundwood nor do they incorporate price interactions between consumption and production (Vincent 1991). The Global Trade Model, developed by the International Institute of Applied System Analysis (IIASA) (described fully in Kallio et al. [1987]), includes price forecasts and predicts a slower increase in consumption of roundwood than the FAO projection; the IIASA base scenario projects doubling of consumption in 50 years, rather than 30. Roundwood prices, forecast through 2010 on a base year of 1980, increase most rapidly, especially for non-coniferous species, almost doubling in most regions, whereas real prices of processed products and coniferous sawnwood remain constant. Vincent (1991) notes that an increase in roundwood production in developing countries, and especially Asian developing countries, is not forecast by the IIASA model that anticipates increases in non-coniferous roundwood supply originating in temperate rather than tropical regions. A major part of this lack of increase is undoubtedly due to the IIASA model's anticipation of an unresponsiveness of long-term roundwood supply to price increases, one of the faults of the model according to critics (Vincent 1991; Cardellichio and Adams 1988, cited in Vincent 1991).

To account for these and other shortcomings, the IIASA model was reworked at the Center for International Trade in Forest Products (CINTRAFOR) (Cardellichio et al. 1989, cited in Vincent 1991). By further disaggregating the regions specified in the IIASA, and re-estimating the supply and demand functions incorporating new theoretical understanding of timber trade flows, the newer version, dubbed the CINTRAFOR Global Trade Model (CGTM), predicts an even lower rate of growth of consumption and trade of most products than the IIASA model. Where the latter predicted price increases of all products sold in Japan at least 1 percent/year (and some at more than 2 percent/year), the CGTM predicted price increases of all products at less than 1 percent/year. At the same time, roundwood supply is rendered more responsive to prices, but the volume of global trade for tropical countries exporting wood and wood products is anticipated to increase at a much less rapid pace, as domestic consumption eliminates the current export surpluses (Vincent 1991).

Finally, a timber trade model that takes a smaller geographic scope is evident in Sedjo and Lyon (1990, cited in Vincent 1991). The Timber Supply Model (TSM) analyzes trade in ten roundwood-producing regions and, uniquely, assumes optimal forest investment over time. The forecasts of this model align most closely with the CGTM, and predict growth of consumption of industrial roundwood to average 0.6 percent/year to 2035, with production in the Asian regions remaining more or less constant (Vincent 1991).

The various perspectives on the future of the tropical forests and hardwood trade form a continuum running from pessimistic to cautiously optimistic, as did the projections and forecasts made 20 years ago, when much less was known about the extent and characteristics of these forests. Some of the discrepancy in views and judgments is due to the orientation of the various reviewers, i.e., whether the focus is on hydrology, biology, timber economics, or other aspects, as well as to the incomplete nature of the data. The 1982 FAO study on the state of the tropical forests was the first attempt to list resources within a uniform system of classification allowing comparison and aggregation of reserves within different countries. However, due to the infrequent inventories of forest in the tropics, "even these data probably only resulted in a modest improvement in the accuracy of estimates of resources and reserves" (Grainger 1986).

While forest boundaries are relatively fixed in temperate countries, land use in much of the tropics is in a relatively early period of transition. Boundaries are often fluid, and forests are generally treated as "common property." Forest lands have competing end-uses. When forests are cleared for agriculture, most of the timber is destroyed.

Logging roads open up previously inaccessible lands to cultivators who then clear more forest, reducing potential second-harvest volumes even more. As domestic markets grow, and as forest resources are depleted of the most valuable species, it would be expected that exports would decline as priority would be given to meeting domestic demand.

The prices of tropical hardwood products vary greatly by species, quality, and country of origin. Due to the large proportion of tropical hardwood in traded hardwood sawlogs, the tropical hardwood log price closely follows the world price for hardwood sawlogs. The real price of tropical hardwood sawnwood stayed constant between 1965 and 1980, but the price of plywood declined significantly, encouraging increased use (Grainger 1986, 19).

CONCLUSION

• Both domestic and international demand for tropical hardwoods has grown rapidly in the past several decades, and demand will continue to grow as national and international economic activity expands.
• The resource base from which tropical hardwoods come is being degraded in most parts of the tropics.
• Government policies and insecurity of tenure create incentives for the forests to be treated as if they were nonrenewable resources, and deforestation and degradation by logging, shifting cultivation, cattle ranching, and other uses of forested land are driven by undervaluation of timber, of non-timber goods, and of environmental services.
• There is a consequent lack of investment to maintain or enhance timber production, so that timber taken from forests is not being replaced by regenerating natural forests or by plantations.
• As the prime forests are depleted of high-value timber, future tropical timber harvests will come from economically more marginal stands. As forest land becomes scarcer, it will also become more valuable for its non-timber goods and environmental services. Whether this increasing social value will promote improved natural forest management and increased forest investment depends critically on the institutional and policy reforms. Were appropriate reforms implemented, it might be possible to manage natural tropical forest for multiple use while also investing profitably in mixed-species plantations.

CHAPTER 3

Natural Forest Management

DUE TO the demonstrably high economic value of tropical moist forests and the heavy loss of environmental services frequently associated with deforestation, countries with significant forest holdings might reasonably be expected to expend considerable effort to ensure the long-term viability and productivity of their forests. In fact, only a very small percentage of the world's productive tropical forest, approximately 4 percent according to one estimate (Lanly 1982), is subject to some level of harvesting regulation and/or silvicultural treatment designed to promote the regeneration of desired tree species. According to another estimate (Poore et al. 1989), less than 1 million hectares of 828 million hectares of remaining productive tropical forests in 1985 is demonstrably under sustained-yield management.

This widespread absence of control over forest exploitation is difficult to attribute to a lack of technical information concerning viable means of managing tropical forests (Leslie 1987). Models of natural forest management have been developed for a wide range of tropical forest types, although their applicability to other geographic regions and long-term sustainability is often poorly understood. While it is economic, social, and institutional constraints on forest investment that constitute the principal obstacles to natural forest management (Wyatt-Smith 1987b; Vincent et al. 1987; Poore et al. 1989), an understanding of the ecological basis and silvicultural characteristics of forest management systems is essential to know how these constraints might be modified to facilitate better management.

ECOLOGICAL CONSIDERATIONS

Of the many environmental factors that influence tree growth, light is perhaps of greatest interest to forest managers owing to the relative ease of its manipulation through silvicultural treatment. The highly variable nature of understory light in tropical forests, along with its importance for species replacement processes, has long been recognized (Richards 1952; Bazzaz and Pickett 1980; Whitmore 1984). In studies of forest regeneration, the presence and size of canopy gaps have often been used as an indirect measure of light availability (Hartshorn 1978; Denslow 1980; Brokaw 1982; Becker 1983). This has led to the view that tropical forests are a mosaic of replacement opportunities, or canopy gaps, with light acting as a major factor determining the ability of a given tree species to colonize a particular forest site (Strong 1977; Bazzaz 1983, 1984; Brokaw 1985). Thus, forests consist of a mosaic of patches, sharing a common history of prior disturbance. The groups of trees occupying these patches are known as stands, and are the basic units that silvicultural techniques aim to manipulate.

In a study of the tropical forest of the Solomon Islands, Whitmore (1974) divided the majority of tree species into four groups, or guilds, of species ecologically similar with respect to their reproductive response to gaps. Whitmore's classification provides a useful basis for the recognition of the following guilds of tropical plants for the purpose of forest management in Africa, Latin America, and Asia:

Pioneer Species. The species of this guild require large gaps for both establishment and growth, are fast growing (often greater than 1.5 cm annual diameter increment), and are generally short-lived (15 to 40 years). Their branching is often orthotropic (ascending, and developing in a similar manner to the leading stem), resulting in a spreading crown and a relatively low proportion of wood in the bole. Their wood is very lightweight, and often of low or no market value. Included in this guild are such genera as *Macaranga*, *Musanga*, and *Cecropia*.

Building-Phase Species. Often similar to the pioneer guild in their establishment requirements and growth rates, these species differ from the pioneers primarily in that they live longer and possess stronger apical dominance. Consequently, they are capable of producing a long, straight bole. Their wood is marketable as paper pulp, plywood, and light construction timber. Examples of genera in this guild include *Terminalia*, *Bombax*, and *Periserianthes* (*Albizia*) *falcataria*.

Climax/Light Hardwood Species. This guild includes species that can establish and grow beneath closed canopy, but benefit significantly from gaps of small-to-moderate size. Under optimal growing conditions, diameter increment in these species can approach 1.0 cm/year, but is generally lower. The species of this guild constitute the bulk of the tropical timber trade, being highly valued for plywood, construction timber, and interior joinery. Many of the world's "mahogany" species figure prominently in this guild, including *Swietenia* and *Cedrela* in Latin America, *Khaya* and *Entandrophragma* in Africa, and many Dipterocarpaceae in Asia (e.g., the red meranti *Shorea*). Teak (*Tectona grandis*) is a prominent member of this guild in Asian deciduous forests.

Climax/Heavy Hardwood Species. Species with seedlings able both to establish and to grow to maturity beneath small gaps or perhaps even closed canopy comprise this guild. These trees are slow growing (less than 0.5 cm annual diameter increment) and commonly live for several hundred years. Their wood is dense, fine-grained, and often quite valuable. They are widely sought after for interior joinery, furniture making, and specialty uses. Examples of this group include ebony (*Diospyros*), rosewood (*Dalbergia, Afzelia*), greenheart (*Ocotea*), merbau (*Intsia*), and ironwood (*Eusideroxylon*).

Although the examples given for these guilds are drawn from timber trees, they also include non-timber trees, shrubs, and vines that possess similar regeneration requirements. Because the silvicultural treatment of tropical forests is generally directed toward guilds of trees rather than individual species, practices intended to benefit a particular guild of timber trees, such as the building-phase species, will usually benefit the non-timber species of the guild as well. Hence, proper management for the timber trees of a guild will also constitute proper management for the fruit trees and other plants of that guild. Conversely, short of enrichment planting, it would be difficult to devise silvicultural practices directed toward a single species only, exclusive of the other species in its guild, unless selective thinning is practiced. On the other hand, if timber species are in a different but competing guild, such as mature-phase emergents and main-canopy trees, from those producing other goods, a decision must be made on the relative importance of competing guilds, and the forest managed accordingly. Often, such conflict can be minimized, because the guilds assume differing importance during regeneration after logging. In Sri Lanka, for instance, important non-timber products, including rattan, medicinal plants, and sugar palm, increase end yield early in regeneration, and later are replaced by climax timber species (C. V. S. Gunatillake, personal communication).

For most tropical forests, silvicultural methods of enhancing timber production will necessarily focus on the two middle guilds of trees: the building-phase and the mature-phase/light hardwood species. Only these two groups combine the attributes of marketable wood and fast-to-moderate growth rates, making their management not only technically feasible, but economically promising. The pioneer species are excluded from commercial management because their wood is seldom marketable, although their presence in forests may be desirable for soil protection or as a nurse for commercial tree species regeneration. The fourth guild, the climax/heavy hardwood species, produces highly valuable wood. Generally, these trees grow so slowly that they are excluded from systems requiring even modest capital investment, since the benefits from their eventual harvest must be discounted over an inordinately long time period (100 plus years). However, protected forests such as extractive reserves or critical watersheds, in which the principal products are non-timber goods and forest services, can provide excellent opportunities for realizing long-term benefits from the climax guilds.

The role of the seedling layer in the natural regeneration of many tropical tree species, particularly in the climax guilds that dominate the mature phase of the forest cycle, is of great significance to the management of commercial forests. For example, although the dipterocarps of Asia generally require at least a small canopy opening to grow to adult size, the probability of seeds being placed in such an opening is low due to poor seed dispersal and infrequent fruiting (Ashton 1982). Combined with a lack of seed dormancy (Ng 1980), such a low probability places considerable importance on the ability of dipterocarp seedlings to persist in the low-light conditions of the forest understory until a regeneration opportunity arises. Such an opportunity can result from a canopy gap due to natural tree mortality in an undisturbed forest, or by selective logging in a commercial forest. The loss of the seedling.layer, however, can have potentially enormous consequences for the future species composition of the forest. A similar dependence on seedling populations in the regeneration of many of the African mahogany species has been noted (Asabere 1987; Kio 1987).

The difficulty of characterizing light availability beneath closed canopy (Chazdon and Fetcher 1984), and perhaps also the high visibility of gaps, means that much of the research concerning natural forest regeneration has focused on the establishment and growth of light-demanding species of the pioneer and building-phase guilds (notable exceptions include Bjorkman and Ludlow 1972; Pearcy 1983; and Chazdon 1985). This approach to studying forest regeneration has been applied to Mesoamerican forests in particular, where canopy turnover rates are high (Hartshorn 1980; Hubbell and Foster 1983), and a substantial proportion of the canopy species are considered to require gaps for regeneration (Hartshorn 1980; Garwood 1983).

However, a significant number of tropical tree species rely on seedling persistence beneath closed canopy as an essential component of their regeneration (Yap 1982; Whitmore 1984). The details of this pattern of regeneration are poorly understood, especially with respect to physiological ecology (Whitmore 1984). In some groups, such as the dipterocarps of Southeast Asia, rapid photosynthetic adjustment to changes in understory light availability appears to play a critical role in seedling survival and growth (A. Moad, personal observation, 1987). Additional research in this area, particularly if conducted within the context of field trials of various management systems, could contribute significantly to the development of silvicultural methods that enhance the regeneration and productivity of tropical forests.

MANAGEMENT SYSTEMS

A wide variety of silvicultural methods has been developed for the long-term management of tropical moist forests. Comprehensive reviews of silvicultural systems for different tropical forest types (Wyatt-Smith 1987a; Wadsworth 1987; Schmidt 1987), and the technologies necessary for their implementation (OTA 1984), have recently been published. Broadly speaking, forest management systems can be characterized as either monocyclic or polycyclic. Monocyclic systems, also known as shelterwood systems, attempt to produce a fairly uniform crop of trees through intensive harvesting and/or extensive silvicultural treatment. Polycyclic systems, on the other hand, aim for a less uniform, mixed-age stand of timber trees through selective felling, occasionally enhanced by limited silvicultural treatment. In monocyclic systems a single, comprehensive harvest of all marketable stems is envisioned at the end of the growth cycle of the forest. In polycyclic systems, two or more harvests are anticipated during the same period of time, with each harvest being less intensive than a single, monocyclic harvest.

Whichever the system, the goal, as in temperate hardwood forests, has been to maximize sustainable timber yield, that is, to attempt to generate as constant and predictable a flow of timber in as high a volume as the potentialities of site and forest will allow. It will surprise no one that this has yet to be achieved. Historically, too, one or another system, devised through analysis of individual stands, has generally been applied uniformly over whole forests, regardless of site conditions or variation in timber or non-timber stocking. This has not only proven ecologically suboptimal (Ashton 1992; Ashton et al., in press), or often unachievable, but is inimicable to multipurpose management (Bowes and Krutilla 1989).

Although often varying substantially in detail due to local ecological and socioeconomic conditions, the silvicultural practices developed for the management of moist tropical forest fall into the following broad categories.

Uniform Shelterwood Systems

Applicable primarily to forests in which the post-logging stocking of desirable seedlings is high or the rapid establishment of seedlings is likely, uniform shelterwood systems are designed to produce an even-aged stand of building-phase or mature-phase/light hardwood trees for harvest on a monocyclic felling basis (OTA 1984). The principal feature of this system is a uniform reduction in the forest canopy cover, either through intensive harvesting or by poison-girdling of non-desired species. The intent is to open up the understory to increased light penetration, thereby promoting rapid growth in the seedlings and saplings of desired species.

Perhaps the most carefully designed version of the shelterwood system is the Malayan Uniform System (MUS), developed for the lowland dipterocarp forests of the everwet lowlands of the Malay Peninsula (Wyatt-Smith 1963). In this system, a selectively logged forest with adequate seedling stocking was treated by poison-girdling virtually all trees of non-desired species, and those trees of desired species of greater than 30 cm diameter. In this way the canopy was opened up gradually as girdled trees died and disintegrated while standing, without incurring the additional damage to understory vegetation that harvesting non-desired trees would entail. Harvesting of the resulting stand, expected to be both even-aged and richer in marketable trees than the original forest, was projected on a 70-year rotation (Wyatt-Smith 1987a).

The MUS worked well in lowland forests with high dipterocarp seedling densities, but was in conflict with felling expectations on sites with low densities, since it recommended deferment of logging until minimum stocking requirements were met (Ismail 1966). Moreover, as logging progressed to hill forests, where seedling stocking is generally low and unevenly distributed, the MUS was found to be inapplicable (Burgess 1968). This, combined with the conversion of most lowland forests to agriculture, led to the general abandonment of the MUS in favor of the Selective Management System, a polycyclic system based on advanced (pole-size) regeneration (Tang 1974).

It seems probable that many of the forest areas managed under earlier versions of the MUS would be producing second-rotation timber harvests by now had they not been converted to agriculture (Jabil

1983; Leslie 1987). However, the extensive poison-girdling of noncommercial trees required by the MUS involves a number of potentially serious problems, including high labor costs, increased erosion and nutrient leaching due to canopy removal, and the removal of trees from the forest that may be marketable in the future (OTA 1984), not to mention reduction or elimination of species producing non-timber goods that may have consequent sociopolitical consequences.

A Tropical Shelterwood System (TSS) similar to the MUS was tested extensively in seasonal evergreen forests in several African countries, where large tracts of forest were cleared of canopy cover in an attempt to develop even-aged stands of commercially valuable climax species (Kio 1979). During the late 1940s and early 1950s, as many as 200,000 hectares were treated under the TSS, involving extensive climber cutting and poison-girdling of noncommercial trees prior to selective felling of marketable trees. Existing regeneration was considered adequate at stocking rates in excess of 100 seedlings per hectare of desired species, mostly in the Meliaceae (Lowe 1978). The results were disappointing, however, due to a general failure to secure the regeneration of desired species and the high labor requirements of weed control (Asabere 1987; Kio and Ekwebelam 1987). As a result, shelterwood systems are not currently being used to any significant extent in Africa (Kio and Ekwebelam 1987). Uniform shelterwood systems for selectively logged forest have been proposed for use in Latin American forests (Wadsworth 1987), but to date are largely untested there (Wyatt-Smith 1987a).

Strip Shelterwood Systems

In addition to the monocyclic systems described above, shelterwood systems have been designed to permit the harvest of limited proportions of forest on a polycyclic basis. Jordan (1982) has proposed a system for the Amazon basin in which strips of forest are intensively harvested on a rotating basis, allowing nutrients, washed downslope from a cleared strip, to be captured by a contiguous forested strip. A similar system is currently being tested in moist premontane forests in the Palcazu Valley of eastern Peru, in which strips 20 to 40 meters wide are clear-cut to simulate natural gap formation processes (Tosi 1982). All wood is removed by nonmechanical means to minimize soil disturbance, with sawn timber, poles, and charcoal comprising the principal products. The cleared strips, which are not burned, are then allowed to regenerate through stump sprouting and seeding-in of pioneer and light-demanding climax species from the surrounding undisturbed forest, much as might occur in a natural forest gap. An inventory of regeneration conducted fifteen months after the clear-cutting of experimental strips in 1985 showed high seedling stocking, with twice the original diversity of tree species colonizing the strips (Hartshorn et al. 1987). Relogging of the strips is projected on a rotation of every 30 to 40 years, but current growth rates suggest that actual rotations may be shorter (G. Hartshorn, personal communication, 1987). Although probably not applicable on a commercial scale using heavy machinery, such strip shelterwood systems have considerable promise for community forests and small commercial operations using small machinery or draft animals.

An economic analysis of the Peruvian project indicates a net profit of $27,500 per hectare logged, based on a gross income of $57,600 from all wood products produced, which are processed locally (Hartshorn et al. 1987). Given a rotation of 30 years, this yields an averaged income of $917/hectare/year for the first cycle of the logging rotation. Subsequent income levels will depend on the volume and quality of the wood produced, which are expected to approach that of the original forest, as well as changes in demand, which is expected to rise (G. Hartshorn, personal communication, 1987). In order to provide a constant supply of wood to the local sawmill and steady employment for local communities, the project will be managed on a sustained-yield basis, with annual rates of logging set at one-thirtieth of the total productive forest area (Hartshorn et al. 1987). It is important to note that legal title to the forest has been given to the local communities via a cooperative corporation, ensuring both local support for the project and long-term tenure of forest lands (Tosi 1982).

A number of shelterwood systems were tested in seasonal evergreen forests in Nigeria by J. D. Kennedy in the 1930s in a comprehensive set of experiments designed to enhance natural regeneration under a variety of forest management methods (Lowe and Ugbechie 1975). Among the systems tested was the Walsh System, in which several 12-hectare compartments were clear-cut and burned, leaving only a few mature mahogany trees for seed production. Regeneration was allowed to proceed by means of seed from these and nearby trees, with some thinning of non-desired seedlings and climber cutting performed to reduce weed competition. When last assessed, in 1970, the resulting second-growth forest was substantially poorer in preferred mahogany species than both the surrounding undisturbed forest and the other shelterwood systems tried (Lowe and Ugbechie 1975). However, the total basal area of all species was higher than that of any other treatment (35 square meters/hectare versus 25 square meters/hectare for uniform shelterwood), much of which consisted of fast-growing light hardwood trees that are less preferred than mahoganies but are nevertheless salable under current market conditions (Kio and Ekwebelam 1987). It seems possible that modification of the Walsh System to conform more with the Peruvian model (strip rather than block clearing, no burning of slash) might encourage the regeneration of both marketable light hardwood trees and the more shade-tolerant climax mahoganies.

The success of strip shelterwood systems depends on the natural establishment and growth of desired tree species on cleared sites. Given the relatively short rotations of most strip systems, this requires the presence of species that are capable of rapid establishment in forest gaps (by stump sprouting, dormant seed, or frequent fruiting), exhibit high growth rates, and produce marketable wood. Early pioneer trees, such as *Macaranga* and *Cecropia*, establish and grow very rapidly on cleared sites but produce unacceptably inferior wood. Certain species of the building-phase pioneer guild, which establish and grow relatively quickly while producing marketable wood, constitute the principal management target of strip shelterwood systems. In contrast, most uniform shelterwood systems are directed toward the climax/light hardwood guild.

The widespread presence of taxa such as *Terminalia* and *Sterculia* in Africa (Kio and Ekwebelam 1987) and *Mimosaceae* in Amazonia (Hartshorn et al. 1987) suggests that strip systems hold potential for these regions. The use of a shelterwood system in which approximately two-thirds of the existing stand is removed in strips 100 to 200 meters wide was recommended for Philippine dipterocarp forests in the early part of this century (Brown and Mathews 1914). However, the potential for the natural establishment of commercial species on cleared sites is uncertain for Asian forests owing to a relative scarcity there of tree species with the necessary combination of rapid regeneration from seed or stump sprouting following logging and marketable wood (P. Ashton, unpublished data). In particular the irregular fruiting characteristics and lack of seed dormancy in virtually all Asian timber trees, including the Dipterocarpaceae (Whitmore 1984), make it unlikely that these species would quickly occupy cleared strips from which the existing seedling stock was removed. For example, dipterocarps and other timber species have been found to be slow to colonize forest clearings in Malaysia and Sri Lanka, in contrast with the rapid invasion of pioneer and building-phase species from contiguous forest (Kochummen and Ng 1977; Gunatilleke and Ashton 1987). This suggests that without more intensive management of regeneration, such as the direct planting of seedlings, strip shelterwood systems similar to those developed in Africa and Amazonia would be unlikely to succeed in the prevailing evergreen forests of Asia.

Selection Systems

Most tropical forests currently managed for timber production are logged on a variation of a selection system, in which a limited number of stems are harvested on a polycyclic basis (Wyatt-Smith 1987a). In India, some 3.6 million hectares of tropical, seasonal forest have long been managed on a selection-cutting cycle of 15 to 30 years (Wadsworth 1987). The dipterocarp forests of Malaysia, the Philippines, and Indonesia are now in theory generally logged on a bicyclic basis, with an anticipated cutting cycle of approximately 35 years and a rotation of 70 years (Schmidt 1987; Uebelhör et al. 1990). Called the Selective Management System (SMS) in Malaysia and the Selective Logging System in the Philippines and Indonesia, it relies on the advanced regeneration of climax/light hardwood species to form the subsequent timber crop, rather than on seedling growth, as in shelterwood systems. Harvests are therefore more frequent than in shelterwood systems, although often with less timber extracted per harvest.

Liberation thinning, also known as improvement thinning, has been used in conjunction with polycyclic systems in a number of forests to increase growth rates in advanced regeneration (Hutchinson 1987a; Jonkers and Hendrison 1987). This process differs from shelterwood systems in its more limited reduction of the canopy, and in its goal of a mixed-aged timber stand (OTA 1984). Liberation thinning is generally less labor intensive than thinning in most shelterwood systems, although it may require a higher level of training on the part of field crews and staff, since it requires considerable judgment regarding which trees to thin (Hutchinson 1987b).

Liberation thinning has received perhaps the greatest degree of attention and experimentation in Sarawak, Malaysia. Following a pre-felling study of stand composition and structure in selected lowland dipterocarp forest in 1974–75, experiments were designed to test the effectiveness of various types of liberation thinning for enhancing the growth of desired tree species in logged forest (Hutchinson 1979). Field experiments using different intensities and methods of liberation thinning were subsequently carried out in several forest reserves throughout Sarawak during 1974–80 (Hutchinson 1982). Analysis of these field trials resulted in the formulation of Sarawak's current prescription for liberation thinning of selectively logged dipterocarp forests. Briefly, this prescription consists of identifying the single best pole-size tree of a desired species in each 100-square-meter section of forest, determining if that tree is overtopped or otherwise suppressed by noncommercial trees, and poison-girdling the competitors where appropriate

(Hutchinson 1979). In situations where young trees of desired species are either absent or not overtopped by competitors, thinning is not required, thus reducing labor costs while maintaining canopy protection of the understory vegetation and soil (Hutchinson 1987b).

The effects of liberation thinning on tree diameter growth rates are presented in table 3.1. Mean annual diameter increments of trees of various quality classes are shown for logged, untreated forest and for four intensities of liberation thinning two years after treatment: (1) overstory thinning alone, in which all trees exceeding 60 cm diameter at breast height over bark (dbhob) and noncommercial trees exceeding 50 cm dbhob are poisoned; and poison-girdling of trees competing with desired "reserve" trees of minimum diameter requirements set at (2) 20 cm; (3) 15 cm; and (4) 10 cm, respectively. The mean number of trees reserved for liberation was 21.3 per hectare, while the mean number of poisoned trees per hectare was 57 (Hutchinson 1979). As shown in table 3.1, liberation thinning is capable of increasing the average annual diameter increment of commercially desirable species, including most dipterocarps, by 82 to 127 percent, depending on the minimum diameters of the trees reserved. For the liberated trees, the increased growth rates generally approach those of trees located in prime, noncompetitive growth sites (0.83 to 0.96 cm dbh/year) as shown in table 3.2. This suggests that liberation thinning succeeds in creating pockets of favorable conditions around the targeted trees

Table 3.1
ANNUAL MEAN DIAMETER INCREMENT BY WOOD QUALITY GROUP FOR A
REPRESENTATIVE LIBERATION THINNING PLOT IN SARAWAK, MALAYSIA
(TWO YEARS AFTER TREATMENT)
(TREES 10–59 CM DBHOB)

	Type of Treatment				
Wood Quality Group	Control	Overstory Removal	20+ cm	15+ cm	10+ cm
Commercially desirable	0.33	0.55	0.60	0.71	0.75
Commercially acceptable	0.17	0.33	0.35	0.45	0.45
Others, grown to timber size	0.20	0.36	0.46	0.50	0.50
Not botanically identified	0.17	0.31	0.38	0.49	0.49
Others, somewhat small	0.14	0.30	0.42	0.42	0.53
"Weed" species, shade tolerant	0.20	0.24	0.35	0.31	0.31
"Weed" species, light demanding	0.43	0.46	0.85	0.49	0.93

SOURCE: Hutchinson 1979.

Table 3.2

ANNUAL MEAN DIAMETER INCREMENT BY CROWN ILLUMINATION CLASS FOR A
REPRESENTATIVE LIBERATION THINNING PLOT IN SARAWAK, MALAYSIA
(TWO YEARS AFTER TREATMENT)
(TREES 10–59 CM DBHOB)

	Crown Illumination Class*				
Treatment	*I*	*II*	*III*	*IV*	*V*
Control, no treatment	0.96	0.59	0.31	0.21	0.12
Overstory removal	0.81	0.81	0.57	0.35	0.74
Liberation thinning, 15+ cm	0.73	0.96	0.68	0.54	0.24
Liberation thinning, 10+ cm	0.81	0.89	0.60	0.56	0.40

SOURCE: Hutchinson 1979.

* I = Emergent Crown
II = Full Overhead Light
III = Some Overhead Light
IV = Mostly Side Light
V = No Direct Light

(Korsgaard 1986). It is not yet known, however, for how long these increased growth rates are sustained.

From a forest management standpoint, the priciple benefit of the Sarawak system of liberation thinning is a reduction in the time required for the second harvest. The exact extent of this reduction will depend on site conditions as well as continued high growth rates, but could be as much as 33 percent, or from 45 to 30 years (S. Korsgaard, personal communication, 1987). The labor requirement for thinning was determined to be 3.1 to 4.2 person-days/hectare for treatment of reserved trees down to 10 cm diameter (Hutchinson 1982). However, a thorough economic analysis of the Sarawak liberation thinning program, based on all costs and projected benefits, has not been conducted (Korsgaard 1986). Liberation thinning is currently being applied to approximately 5,000 hectares of logged forest per year in Sarawak, or about 4 percent of the total area logged annually (E. Chai, personal commnication, 1987). Additional Sarawak Forestry Department staff is being trained to extend liberation thinning to larger forest areas.

Liberation thinning has also been tried, with varying levels of success, in the forests of tropical America and Africa. In Surinam, the Celos Silvicultural System (CSS) incorporates a series of forest refinement thinnings in a polycyclic management system (de Graf 1986). The thinning treatment used is less selective than that of Sarawak, with the initial thinning encompassing all trees above 20 cm, or an average of 73 poison-girdled trees per hectare. A second thinning treatment is anticipated in 8 to 10 years following the first, and possibly a third as well, although the required intensity of thinning is as yet undetermined (Jonkers and Hendrison 1987). The annual diameter increment of the remaining commercial trees is expected to increase from 0.4 to 1.0 cm, resulting in a yield of 40 cubic meters/hectare of harvestable volume in 20 years (Schmidt 1987). The labor cost of the initial treatment is 2.8 person-days/hectare, while the cost of subsequent treatments will depend on their intensity (Jonkers and Hendrison 1987). The lower labor requirement of the CSS thinning treatment in comparison with liberation thinning in Sarawak can be explained by the less site-specific nature of thinning prescriptions in the former (de Graf 1986).

Light thinning of noncommercial trees has been tried in several African nations in conjunction with selection systems (Nwoboshi 1987). In Gabon, for example, approximately 1 million hectares of forest received light thinning in the 1950s to promote the establishment and growth of *Aucoumea klaineana*. Thousands of hectares were managed in a similar fashion in Zaire, Nigeria, Ghana, and the Ivory Coast (Schmidt 1987). In virtually all of these cases forest improvement thinning has been abandoned, in part for silvicultural reasons, such as the stimulation of climber and weed competition, but primarily due to institutional difficulties and lack of funding (Asabere 1987; Kio and Ekwebelam 1987). Perhaps the most systematic study of the effects of thinning operations in Africa was carried out in the Ivory Coast in the 1970s by the Société Ivoirienne de Développement des Plantations Forestières (SODEFOR) and the Centre Technique Forestier Tropical (CTFT) (Schmidt 1987). Two intensities of thinning (35 percent and 45 percent of total basal area) were tested for enhancement of growth in 10 plus cm diameter commercial trees, with the results showing a 50 to 75 percent mean increase in volume increment in the 73 target species. The 10,000-hectare Yapo forest is currently being managed with the SODEFOR-CTFT system on an experimental basis to evaluate its potential for widespread application (Schmidt 1987).

As indicated above, the first harvest in a selection system is expected to leave a stand of pole-size trees and somewhat relieve competition for light, water, and nutrients, and thus enable relatively rapid growth to harvestable size. In fact, however, advanced regeneration is often sparse or severely reduced by logging damage, and the growth response of the remaining trees less than expected (Appanah and Salleh 1987; Wyatt-Smith 1987b; Uebelhör et al. 1990). Originally intended as an alternative to the shelterwood system in areas where advanced regeneration is significant, selection systems have often been adopted to reduce cutting cycles for short-term economic expediency, irrespective of forest regeneration patterns (Tang 1974). As Bowes and Krutilla

(1989) have poignantly observed, though, the demands on forest change too frequently, both quantitatively and qualitatively, for long rotation systems to be economically attractive. The challenge must be to develop shorter rotation systems, therefore polycyclic, that are ecologically achievable. As indicated by the experience of researchers in Sarawak, Surinam, and the Ivory Coast, however, selection systems can be formulated that show promise for sustained timber production, provided that the systems are properly applied.

Liberation thinning has never to our knowledge been applied to species producing non-timber goods as well as timber species. Where stocking of the latter is adequate, this could result in direct competition between the two for space. There is some evidence that the maintenance of species diversity enhances performance (Hubbell 1990). In those many forests where stocking regeneration of timber species is suboptimal, then the encouragement of non-timber products could add significant value.

Ensuring Regeneration

With the exception of strip shelterwood on cleared sites, the forest management systems described above depend on adequate stocking of seedlings, saplings, or pole-sized trees for their successful implementation. Since relatively few tropical timber species exhibit extensive seed dormancy or patterns of continuous or frequent fruiting (Whitmore 1984; OTA 1984), the absence of natural regeneration or its destruction during logging virtually negates the possibility of silvicultural management for future production (Wyatt-Smith 1987a). The abandonment of silvicultural practices in Africa has more often been due to a lack of regeneration with which to work than the inadequacy of prescribed methods for enhancing tree growth (Asabere 1987; Kio and Ekwebelam 1987). Even in Sarawak, where the selection system practiced is considered one of the most successful (Schmidt 1987), the effectiveness of liberation thinning is constrained by inadequate stocking of pole-size reserve trees, generally present at 30 trees/hectare rather than the 100 trees/hectare desired (Hutchinson 1979).

In some cases, poor regeneration of commercial species is a natural feature of the forest being managed (Nwoboshi 1987). Equally often, however, it is because of the widespread destruction of young trees and understory seedlings during logging operations (Myers 1980; OTA 1984; Wyatt-Smith 1987b). Accidental damage to smaller tree stems during felling, severe soil compaction of 4 percent or more of the logged area by roads, and the destruction of as much as 40 percent of the understory seedling layer by log extraction all contribute to the problem of inadequate stocking of commercial species following harvesting (Marn and Jonkers 1981; FAO 1981).

Most countries with tropical moist forests have stringent policies concerning allowable logging damage, but enforcement is often confined to minimum diameter felling limits owing to manpower constraints, weak concession agreements and, in some cases, collusion between concessionaires and enforcement officials (Leslie 1980; Wyatt-Smith 1987a; Uebelhör et al. 1990). Particularly damaging is the practice of repeat logging, in which second and even third concessions are given out for the same forest in the space of only a few years. This practice results in considerably more logging damage than would occur with a single, more intensive harvest (OTA 1984). The use of enrichment planting of seedlings may be possible in some of these damaged forests, but it is unlikely to constitute an economically attractive alternative to natural regeneration in situations where such regeneration is present and can be preserved.

A promising approach, therefore, would be to limit the impact of logging on understory vegetation and soil erosion, particularly in areas where natural regeneration depends primarily on the existing reservoir of seedlings and saplings. Although the potentially devastating impact of unrestrained logging practices is well known, the economic costs and feasibility of more stringent logging control are relatively poorly understood (see chapter 8). Studies exist, however, indicating that modest control of logging operations can substantially reduce damage to regeneration at moderate or even no additional cost. In one study of controlling logging damage in Sarawak, it was found that the use of directional felling and planned skid rails reduced mortality to the remaining tree stand by 33 percent and understory destruction by 40 percent, while increasing skidding efficiency by 36 percent, thereby reducing harvesting costs by as much as 23 percent (Marn and Jonkers 1981). The use of directional felling and skidtrail planning was found to produce similar results in the CCS project in Surinam (Jonkers and Hendrison 1987—see chapter 8).

Given both the importance of the regeneration stock to future production and the feasibility of its preservation during logging, the first objective of any silvicultural system of forest management should be the control of logging damage. In many countries, control would primarily involve the enforcement of existing harvesting regulations; in other countries, new regulations might be required. Without such enforcement, it seems unlikely that forest managers will be either institutionally or practically capable of implementing the more demanding silvicultural systems outlined above.

FAILURE TO IMPLEMENT MANAGEMENT PRACTICES

Despite the wide variety of silvicultural systems available, few of the world's tropical forests are currently being managed on a systematic basis (Leslie 1987; Wyatt-Smith 1987a). In a survey of 76 countries with tropical forests, it was found that only 42 million hectares, or 20 percent of the 210 million hectares that have already been logged worldwide, are being managed even in theory (Lanly 1982). As indicated earlier, Poore et al. (1989), in the most careful analysis to date, concluded that less than 1 million hectares out of the 828 million within timber-producing members of ITTO, were demonstrably under sustained yield management in 1985. These countries contain more than 70 percent of the world's tropical forests. In Asia, where the majority of managed tropical forests are found, about 19 percent of the region's total productive, closed broadleaf forest is being intensively managed, implying some degree of harvesting control and/or subsequent silvicultural treatment (FAO 1981).

In many cases this lack of forest management represents the de facto forestry policy in countries where governments have recognized the economic imperative to respond to current market and development priorities rather than place faith in seemingly uncertain future demand and yields (see Bowes and Krutilla [1989] for a temperate forest example). In other cases, the lack of management reflects conditions where forestry regulations are inadequately defined and where governments have made the implicit decision to "mine" forest resources on a nonrenewable basis as a source of immediate, temporary income (Burns 1986; Leslie 1980; Uebelhör et al. 1990). Alternatively, the lack of active forest management may reflect the disparity between forest policies and the resources devoted to their implementation. In many countries the rate of logging is simply too great for forestry agencies to deal with effectively, given the constraints on available manpower and finances (Wyatt-Smith 1987b). Finally, a "minimum intervention" approach to the management of natural forests following logging may be preferred in situations where adequate regeneration of desirable tree species is assured or the economic return from silvicultural treatment is in serious doubt (Tang 1987). Forest areas managed in this way represent a low input/low output resource, with protection of the forest justified primarily on the basis of secondary services or the lack of a cost-effective alternative, rather than on timber production (Leslie 1977). Although some forests undoubtedly fit this category, it should be noted that most forests are managed on a minimum-intervention basis more by default than by policy design.

Thus, although most countries possessing tropical moist forests have declared policies for natural forest management, few effectively employ prescribed silvicultural practices (NRC 1982; Wyatt-Smith 1987b). More commonly, the implementation of management prescriptions is often severely behind schedule or abandoned altogether. The principal reasons suggested for the inability of forestry agencies to implement declared management policies include the following:

1. continued conversion of forests to agriculture or rangeland, often in spite of stated policy, making long-term forestry investments unattractive (Myers 1980; NRC 1982);
2. intense political and economic pressure on forest managers to allow practices such as accelerated felling cycles, clear felling, immediate second and third cuts, and leniency with regard to logging damage, which generate short-term income at the expense of long-term productivity (Leslie 1980; Schmidt 1987);
3. unwillingness of governments and private landholders to invest scarce capital in a resource for which the return is perceived to be both unacceptably long-term and uncertain (Leslie 1977; Burns 1986);
4. failure, in some cases, of silvicultural practices to enhance the regeneration and growth of desired species (Kio 1979; Asabere 1987); and
5. the sheer size and ecological complexity of the tropical forests requiring management in relation to the resources available to the agencies assigned to manage them (FAO 1981; Wyatt-Smith 1987b).

These impediments to long-term investment in forest resources can be laid to one or more of the following basic problems in land use and forest policy. Some of them, such as rapid population growth and agricultural land tenure, are outside the normal scope of forest policy but nevertheless warrant mention because of their considerable influence on the use of forest lands. Others, dealing more directly with forest management, are discussed in greater detail in later chapters of this study. These basic problems include:

1. Forest tenure uncertainty
   • Many countries lack an overall land-use policy designating permanent forest reserves, leading to widespread conversion of forest lands.

- Increasing population pressure and inequitable distribution of agricultural land continues to force landless farmers onto marginal forested lands.
- Logging concessions are almost always of shorter duration than forest rotation periods, and guarantees of future rights to production on public land are generally inadequate. There is thus little incentive for private investment in forest protection and management.
- Logging concessions are often awarded or withdrawn on a political basis, making their length of tenure uncertain. Extensive areas of forest are therefore quickly high-graded for the most valuable timber, leaving less preferred yet marketable species for subsequent re-logging.

2. Undervaluation of forest resources
   - Timber values based primarily on extraction cost do not reflect the replacement value of timber.
   - Current methods of assessing forest value are generally based on timber production alone, and thus do not reflect the full value of forests. The two most common omissions are (a) failure to include other forest products, such as rattan, game, fruits, resin, latex, and medicinals; and (b) failure to include environmental/social services, including watershed protection, soil conservation, gene conservation, climate regulation, and recreation/tourism.
   - Low royalty fees and stumpage taxes, while providing windfall profits to logging companies, reinforce government perceptions of low forest value.

3. The distribution of the costs and benefits resulting from logging and silvicultural treatment is seldom taken into account in management decisions, leading to misallocation of resources as well as market inefficiencies.

4. Forest management is focused almost exclusively on timber production, often to the detriment of other forest products and forest services.

5. Inadequate institutional arrangements
   - Widespread political control over the award of timber concessions, felling cycles, allowable logging damage, and silvicultural treatment of logged forests results in public forests being managed primarily for short-term private gain.
   - Inadequate representation of local interests in the formulation of forest policy leads to the undervaluation of forest services and non-timber products, lack of local support for forest protection

and, increasingly, open conflict over forest use, resulting in prohibitively high management costs.

- Conflicts between government agencies (e.g., agriculture versus forestry) often result in irrational land use policies and uncertainty over forest tenure.
- Inadequate knowledge of field conditions on the part of senior forest managers and policy makers, reinforced by the division of forestry departments into field and office personnel, makes site-specific silvicultural treatment difficult.

CONSIDERATIONS FOR FUTURE MULTIPLE-USE MANAGEMENT

The thesis of this book is that economic valuation of the totality of a forest resource, difficult though that may be, provides the soundest available basis to define management objectives. Besides markets and geographical issues such as accessibility and the resources of individual forests, the characteristics and relative areas of their component stands and their site conditions will impose potentialities and constraints on management options. Increasingly, management of natural temperate forests, using similar criteria for defining objectives, entails division of forested landscapes into units to which differing objectives and methods are allocated. Bowes and Krutilla (1989) point out, for instance, that management for maximizing forest value relative to water conservation or recreation and aesthetics, as well as timber production, will be achieved with longer felling rotations than management for timber alone. This implies that differing silvicultural systems will be favored. They also show that the interactions between different management units can increase the value of the whole forest, for recreation or conservation. For instance, tracts that may contain the greatest value when strictly preserved will particularly benefit from being set in larger forest areas allocated to other uses (see chapter 10). Concurrently, diversified management of forested landscapes increases the options for responding to future changes in demand.

It has to be said that few forest services are even beginning to prepare for the level of silvicultural and management skills that are required to manage tropical forests for the future. Particularly, a cadre of field personnel, trained beyond the established protocols to sensitively evaluate conditions on the ground, to plan parsimonious means to collect appropriate data, and to actively lead field labor with insight and enthusiasm, is the keystone for success.

CONCLUSION

The silvicultural management of tropical forests is best considered from the standpoint of ecological guilds of species, which share similar regeneration requirements, rather than individual species. Most forest management systems are directed toward the building-phase pioneer and climax/light hardwood guilds, recognizable in the dynamic cycle of the natural forest, with light being the principal environmental factor that can be manipulated. Apart from enrichment planting, management techniques for non-timber species are consistent with management techniques for timber production: proper management for one group will generally constitute proper management for the other as well, at least when they share the same ecological guild. Notably, liberation thinning that allows for improved performance of species producing non-timber goods as well as timber species may add significant value to those many logged forests where stocking of the latter is suboptimal. Even when timber and non-timber products occupy different dynamic guilds, the two may be accommodated with minimal competition if they grow and yield at different times during stand development following logging.

The existing methods of tropical forest management can be broadly divided into monocyclic and polycyclic systems, of which the uniform shelterwood system and the selective system, respectively, are the most common examples. In general, the tropical forests of Asia have received the greatest level of silvicultural experimentation and application, as well as the highest rates of logging. A number of silvicultural methods have been tested in Africa as well, although less extensively in recent years. Silvicultural methods have been least tested and applied in tropical America, although some models do exist. The technical and socioeconomic requirements for the successful transfer of methods developed in one region to another region are poorly understood.

There is little doubt that as the values of non-timber goods and services of tropical forests become fully apparent, optimal management will require division of forested landscapes into blocks that will be managed for different principal objectives, using differing silvicultural protocols. The interdependencies between and among these blocks may well enhance the value of the whole.

Most commercial tree species depend on seedling and sapling availability in the forest understory for their regeneration. Control of logging damage therefore constitutes a fundamental requirement of virtually every silvicultural system designed to enhance long-term forest production.

The large majority of tropical forests receives little or no silvicultural treatment, and often inadequate protection. The causes of this lack of forest management are primarily economic and social in nature, with the undervaluation of non-timber forest resources and inappropriate institutional arrangements playing a major role.

CHAPTER 4

Undervaluation of Tropical Timber

TROPICAL FORESTS are economic assets that provide multiple products, including timber and non-timber goods and services. As assets, forest resources can generate a return for their owners, which is directly attributable to their scarcity. One widely accepted goal for forest management is to maximize the total net benefits (rents) from all possible commodities and services over time—whether or not these benefits are traded in, and priced by, markets. Under secure land ownership this rent goes to the owners of the forest. To manage the timber supply and the forest on a long-run sustained yield basis and maximize its benefits, both the owner and the government should be able to assess the full value of the forest.

Governments, which own over 80 percent of the world's tropical forests, have exhibited a consistent tendency to undervalue tropical forests in at least three ways: (1) by undervaluing timber, as exemplified by low rates of taxation, subsidized logging, and the terms under which they grant logging concessions; (2) by overestimating the benefits from timber industries and the conversion of natural forest to other uses; and (3) by ignoring non-timber goods and other forest services. The purpose of this chapter is to address the first two problems: the undervaluation of timber and the overvaluation of timber industries and forest conversion. The neglect of non-timber goods and forest services will be addressed in the following two chapters, respectively.

The issues raised by the undervaluation of timber illuminate the need for multiple-use management. Although it might appear that the more undervalued timber is, the more attractive non-timber goods and services become, this is not the case. In fact, to the degree that timber is undervalued by owners and governments, the demand for—

61

and supply of—logging concessions is likely to be greater and the enforcement of logging regulations less strict. Moreover, since many non-timber goods and forest services are joint products with timber, or at least share a common resource base, the pervasive undervaluation of timber leads to less investment in their protection.

Governments' undervaluation of tropical timber leads to excessive and wasteful deforestation and discourages investments in silvicultural improvements, protection against fire and disease, and forest regeneration. This is especially true in view of timber often being the major commercial or marketed product of the tropical forest. If timber, a major internationally traded commodity, is consistently undervalued, one can hardly expect "minor," locally consumed non-timber products and intangible environmental services to be taken into account in forest management. The undervaluation of tropical timber contributes significantly to the undervaluation of all products of the tropical forests, which in turn undermines the long-term sustainability of supplies of those products, including tropical timber. This remains true in spite of the fact that undervaluation may temporarily benefit both producing countries (as a source of foreign exchange) and consuming countries (in the form of lower prices).

In order to ensure the long-term sustainability of tropical forest production, benefitting consumers and producers alike over the long run, adequate investments in forest management, regeneration, and planting are necessary. Such investments will not be forthcoming unless timber is priced at its full scarcity value. The full scarcity value of timber is defined as a price that fully covers logging, transportation, and user costs. User costs consist of the foregone future revenues as a result of current logging; these future revenues reflect the rising value of tropical hardwoods as their natural sources become depleted and/or as demand for timber increases over time. As long as tropical timber prices are influenced by the continuing availability of the resource at prices below its scarcity value, forest investments will remain unprofitable, and both capital and land will continue to be shifted to alternative uses.

The misallocation of scarce resources is further compounded by the overvaluation of the benefits from forest-based industries and forest conversion to other uses. Such misallocation is manifested in the continuing availability of protection and subsidies for those activities that increase the private profitability of forest-based industries and forest conversion beyond their social profitability. Both undervaluation of timber and overvaluation of timber industries, therefore, threaten long-term sustainable production of all tropical forest goods and services.

UNDERVALUATION OF TIMBER PRODUCTION

In many cases, forests now publicly owned once benefitted from a long tradition of efficient and sustainable management by local indigenous communities. However, when governments claim ownership of these forests, they often do not act as exclusive and secure owners who fully value their productive assets. Rather, because their ownership claims are not effectively enforced, these resources are not sufficiently protected by the state or by those to whom the responsibility of protection has been delegated by the state. In addition, they are rarely managed so as to obtain the full value of the forest.

Many developing countries follow policies that return to the public coffers only a small fraction of the value of timber harvested. In economic terms, these policies do not capture the full economic rent. Rent is the value of a resource in excess of the costs of obtaining it. For timber, this would be the price per log minus harvest and delivery costs. Governments have at their disposal a variety of mechanisms and policies with which to capture these rents, including royalties and taxes, land rents, and licensing and export policies (Gillis 1980).

The combination of royalties and taxes charged in many countries is well below the rent of the forest. For instance, the government of the Ivory Coast failed to capture rents of up to $40 per cubic meter of timber in the early 1970s (Repetto 1987). Indonesia receives only 50 percent of the rents from log exports and 25 percent of the rents from sawn timber, leaving $700 million in rent to loggers (Gillis 1988a); Ghana collects less than 40 percent, and the Philippines realizes less than 10 percent (Repetto and Gillis 1988; Boado 1988).

The timber booms resulting from this rent-neglecting behavior have been a leading cause of the dramatic deforestation rates in some countries. In Indonesia, for example, concessions had been awarded for 1.4 million hectares more than the total area of production (or "productive") forests in the country by 1983 (Repetto 1987, 95). The Ivory Coast, with the world's highest deforestation rate (7 percent per year), leased over two-thirds of its production forests to concessionaires during the seven-year period between 1965 and 1972 (Repetto, 1987, 94). While timber harvesting is not the only important cause of this rapid deforestation, timber contractors have by now virtually exhausted the more valuable species, and shifting cultivators have moved in to clear the depleted forests (Repetto and Gillis 1988).

The likelihood of concessionaires acquiring excessively large rents is increased by the fact that many governments do not allow competitive bidding for timber projects (Repetto 1987, 94). Further, the common practice of basing royalties and taxes on the volume of timber harvested rather than on the total volume of marketable timber in a stand leads to high grading, in which only the most valuable timber is harvested, leaving little or no incentive to preserve or protect the remaining trees. Even where royalties and taxes are very low, contract uncertainty may lead to the same result. Thus, trees left at the site are tacitly given little or no value and are often damaged. For example, Repetto (1987, 95) reports that 45 to 74 percent of the trees that remain after harvesting in Sabah are either "substantially damaged or destroyed." The problem in Indonesia appears to be much the same.

Some governments, as in Ghana and in Sabah, Malaysia, have demonstrated that workable royalty systems can be devised that will protect less valuable trees from damage.

> In Ghana, a different royalty rate is applied to each of thirty-nine commercial species, and rates are charged per tree harvested rather than per cubic meter. This system has encouraged loggers to harvest a variety of species and to utilize each stem cut as fully as possible. Sabah imposes specific charges that vary considerably by species and suffers only half as much residual tree damage from logging operations than either nearby Sarawak or Indonesia, which have flat royalty rates. (Repetto 1987, 95)

But far more examples of government forest policy exist that encourage overharvesting and waste. In addition to allowing concessionaires extra profits from resource rents, many governments further increase concessionaires' profits by subsidizing or assuming many of the extraction and marketing costs, including road and infrastructure construction and the surveying, marketing, and grading of saleable timber, as well as the environmental damage costs of timber operations (Repetto 1987, 95).

The incentive to harvest at a rapid rate is further exacerbated by government practices that often stipulate certain cutting deadlines and limit the duration of leases to less than an optimal harvest rotation. Governments often do this in order to prevent the stockpiling of leases, but these practices effectively remove incentives for harvesters to encourage future growth of the forest. A significant number of timber leases in Sabah are for ten years, and 5 percent are for as little as one year, while the optimal harvest rotation is substantially longer (Repetto 1987, 95). Licenses in the Philippines used to be issued for only one to four years, but now they are issued for 25 years, with a chance for renewal (Boado 1988).

Stumpage prices, which normally would signal resource scarcity, are thereby suppressed through various government policies. In Peninsular Malaysia and Ghana, the transition from old-growth to second-growth forests has been hampered by government policies that hold increases in timber charges below increases in stumpage values, resulting in increased returns to rent-seeking in the tropical forest and harvests that exceed sustainable yield (Vincent and Binkley 1991). In both cases, royalties were set well below stumpage values, thus discouraging forest investment, along with insecurity of tenure and, in the case of Ghana, macroeconomic policies that were not conducive to productive investment (Vincent and Binkley 1991).

In summary, forestry policies, including those covering the terms of timber concessions, permissible annual harvests and harvest methods, and levels and structures of royalties and fees, reflect the fact that governments in many developing countries undervalue their tropical timber resources. If the loss of benefits from non-timber goods and services, as well as the environmental and social costs of timber exploitation, were factored into government policy, public authorities would not only charge concessionaires a full rent reflecting wood scarcity but they would also add an imputed rent on top of the stumpage value for all the other services that society will be deprived of by conceding its forest to logging.

OVERESTIMATION OF NET BENEFITS FROM TIMBER INDUSTRIES AND FOREST CONVERSION

Not only is tropical timber undervalued by governments and markets but governments have demonstrated the propensity to overvalue the benefits from development of wood-processing industries and from conversion of forest land to non-forest uses such as farming and ranching.

Wood-Processing Industries

Even when governments realize that they are not gaining full rent from timber concessions, they are often willing to forego these revenues in order to obtain other benefits from the timber industry such as jobs, roads and other infrastructure improvements, and increased foreign exchange earnings (Gillis 1980). However, these potential benefits are all too often overestimated and dissipated by underestimating government costs.

One common development strategy is the promotion of domestic wood-processing industries. In order to increase value added and create domestic employment, many log-exporting countries have shifted from exporting raw wood to exporting finished wood products by encouraging local wood-processing industries, i.e., sawmills, plywood, veneer, fiberboard, particle board, pulp, and paper, etc. Various kinds of investment incentives have been offered by governments. These have predictably led to enormous growth in their wood-processing industries. For example, Ghana

> enacted log export bans, exempted plywood and other wood products from export taxes, granted long-term loans to mills at a real interest rate less than the rate of inflation, and granted a 50 percent rebate on income tax liabilities to firms that exported more than 25 percent of their output. By 1982, these policies had created a domestic industry of 95 sawmills, 10 veneer and plywood plants, and 30 wood-processing plants. (Repetto 1987, 96)

Similarly, in Indonesia in 1983,

> the government raised the log export tax rate to 20 percent, but exempted most sawn timber and all plywood. Mills were also exempted from income taxes for periods of five or six years. With these incentives and the impending log export ban, the number of operating or planned sawmills and plymills jumped from 16 in 1977 to 182 in 1983. (Repetto 1987, 96)

It is clear that such timber industries generate local employment and income, but governments that provide incentives for these industries should be aware of the trade-offs involved. Because local timber industries are often too inefficient to compete internationally without subsidies, incentive policies often result in net economic losses for the country.

> Many of the mills built in response to these incentives in Ghana, the Ivory Coast, the Philippines and Indonesia are small and operate inefficiently. Ghana's plymills use 2.2 cubic meters of log input per cubic meter of output. Conversion in Indonesia's and the Philippines' plymills are 2.3:1, and 2.5:1 respectively. These are far below 1.8:1 in Japan and 2:1 in Korea and Taiwan. (Repetto 1987, 96)

Inefficient wood-processing plants not only underutilize forest resources, which in turn reduce rents from forest resources, they also further reduce government revenues. Several countries have reduced or eliminated export taxes for finished wood products, a clear sacrifice of government revenues. The Indonesian government, for example, actually lost $24 for every cubic meter of plywood exported, compared with potential earnings of $53 in tax revenues if the timber were exported as unfinished logs (Repetto 1987, 96). In the Ivory Coast, export taxes for unfinished logs are 25 percent and 45 percent (depending on the quality of logs), while the rates for exporting similar quality plywood are 1 percent and 2 percent. The result is that for every $20 of domestic value added to a cubic meter of mahogany when it is converted into plywood, the government loses $40 in export taxes (Repetto and Gillis 1988, 344). The protection of inefficient domestic wood-processing, by either banning log exports or imposing high export taxes on logs but not on timber products, will indeed create jobs, but often at a very high cost to the nation (Repetto and Gillis 1988). Estimates for Peninsular Malaysia (Vincent 1992) indicate that average annual losses per sawmill job created totaled $6,100 in economic value-added, $16,600 in export earnings, and $34,300 in stumpage value.

Forest Conversion

Conversion of forest lands to other uses such as resettlement, crop cultivation, and livestock production can generally be observed in many forest-rich countries. Again, government incentives have played a significant role in encouraging intrinsically inefficient activities or competing land uses beyond economically rational levels.

Incentives come in various forms, such as infrastructure expenditures that directly subsidize competing activities, grants to settlers, or low-interest loans and tax reductions/exemptions. In many countries in Latin America, and some countries in Asia such as Thailand and the Philippines, property rights to forest land are established by clearing and converting it to another use (Repetto 1988a). Such policies shift the margin of relative profitability from forestry to competing land uses, thereby encouraging more rapid forest conversion.

A notable example of the overvaluation of these alternative land uses occurred with the Indonesian transmigration program. This program entails the resettlement of 5 million people at a cost of $10,000 per family (per capita GNP is only $550). While original plans called for new communities to be created in converted forests, the poor soils of the outer islands have doomed many of the settlements. As the value of the forests becomes better appreciated, the plans now call for most of the settlements to be located in forested areas, where they are expected to be supported by tree-based industries such as rubber and palm oil (Repetto 1987, 98).

In Latin America, government-sponsored incentives to convert forests to pastures for raising cattle have been substantial. The Brazilian program of credits and subsidies totaled over $731 million between 1966 and 1983 (Repetto and Gillis 1988). The extent of deforestation related to cattle ranching is likewise enormous. It has been estimated that in 1980, 72 percent of the areas categorized via satellite monitoring as deforested were being used for pasture (Repetto and Gillis 1988). However, a recent study of cattle ranching in the Brazilian Amazon (see table 4.1) has demonstrated that it is only the government incentives that make such operations profitable (Repetto 1988).

Table 4.1

ECONOMIC AND FINANCIAL ANALYSIS OF GOVERNMENT-ASSISTED
CATTLE RANCHES IN THE BRAZILIAN AMAZON
(IN MILLIONS OF U.S. DOLLARS)

	Net Present Value Investment Outlay	Total Investment Outlay	NPV
Economic Analysis			
Base case	−2,824,000	5,143,700	−0.55
Sensitivity analysis			
Cattle prices assumed doubled	511,380	5,143,700	+0.10
Land prices assumed rising 5%/year more than general inflation rate	−2,300,370	5,143,700	−0.45
Financial Analysis			
Reflecting all investor incentives: tax credits, deductions, and subsidized loans	1,875,400	753,650	+2.49
Sensitivity analysis			
Interest rate subsidies eliminated	849,000	753,650	1.13
Deductibility of losses against other taxable income eliminated	− 658,500	753,650	−0.87

SOURCE: Repetto 1988.

CONCLUSION

Timber production in many developing countries is consistently undervalued by governments. In general, governments do not charge loggers for the use of forest resources based on the economic rents (stumpage value) generated by the timber. Hence, concession fees, license fees, and land rents are far too low. Types and terms of taxes and concession agreements encourage suboptimal rotation rates and inefficient resource use. Moreover, the loss of value of non-timber goods and forest services, such as soil erosion and watershed protection, wilderness, wildlife habitat, and recreation, is rarely incorporated into the costs of timber production. These values can conversely be considered benefits from reforestation. In addition to giving concessionaires extra profits through untaxed resource rents, many governments increase concessionaires' profits on timber from public land by assuming infrastructure costs such as those of construction of roads and port facilities, surveying, and grading.

The failure to capture resource rents, together with the inappropriate structure and duration of concession agreements, results in excessive and wasteful deforestation and the foregoing of substantial government revenues that could be used for reforestation and protection of reserved forests.

While tropical timber has consistently been undervalued, the net benefits of domestic timber-processing industries and of converting forests to other uses have commonly been overestimated. Government activities such as subsidizing competing uses through infrastructure investments, and the provision of low-interest loans and tax reductions/exemptions, have played a significant role in many developing countries in encouraging intrinsically inefficient activities and accelerating forest conversion.

In order to manage forests on a long-run sustained yield basis and to achieve maximum social welfare, governments must fully value their forests. Full valuation of forests includes not only full valuation of timber in both absolute terms and relative to other activities but also, as we shall see in succeeding chapters, full consideration and accounting of non-timber goods and services.

CHAPTER 5

Non-Timber Forest Products: A Major Component of Total Forest Value

IN MOST tropical countries, government forest policy is primarily concerned with utilization of forest resources for the production of timber and, secondarily, fuelwood. For the majority, timber and fuelwood are important as sources of building materials, energy, and industrial raw materials as well as foreign exchange. Tropical forests, however, yield other products, which in some countries may be of comparable or even higher value than timber. These products are generally referred to as "minor" forest products—an unfortunate label, for though not highly visible, their harvesting from the forest is a major activity in the rural economy. "Non-timber" goods, as they are referred to in this study, also currently account for several billion U.S. dollars in world trade annually (de Beer, 1990).

An exhaustive list of non-timber goods could number thousands of products, ranging from exudates (gums, resins, and latex) to canes (rattan); from edible nuts, fruits, and vegetables to fungi and spices; from meat and by-products from game animals, including mammals, fowl, reptiles, fish, and insects, to the animals themselves for the pet, zoo, and tourist trade; from fuelwood and fodder to biochemically active plants, microorganisms, and insects for diverse pharmaceutical and medicinal uses.

These products usually differ from timber in the following respects: (1) the greater variety of products and of species; (2) the shorter frequency of harvest cycles; (3) the smaller yield per unit area in natural forest; (4) the frequently higher monetary value per unit weight; and (5) the propensity for synthesis from other resources or substitutes. Since non-timber goods tend to be less bulky, they require harvesting that is both more labor intensive and generally less destructive than timber exploitation, at a relatively small capital investment.

Although non-timber goods can be found in almost any type of forest, tropical rainforests, as the most species-rich ecosystems, are potentially the richest source. Despite the fact that a large number of these products have already been identified, the number of commercially exploited non-timber goods is almost certainly still a small fraction of the available potential.

The Social and Economic Importance of Non-Timber Products

An assessment of the economic value of non-timber products must include consideration of their utility internationally, regionally, and locally. Some products are marketed, processed, and used far from the forests by industries, urban dwellers, or distant foreign economies. Others enter into the national economy, or are collected and used exclusively by local forest dwellers and adjacent rural populations (Clay 1988; de Beer and McDermott 1989). Considerable overlap exists among these categories because products collected in large quantities, even for export, provide local income and employment through harvesting, while products consumed and processed locally may also have regional or national significance.

Many non-timber goods, however, are not traded in markets. Even those that are traded are rarely reported in national or international (e.g., Food and Agriculture Organization) statistics of the forest product trade (Hamilton 1986, 8; de Beer and McDermott 1989). This lack of reporting strongly implies that the overall value of non-timber products is seriously underestimated. Recent studies in various tropical countries have shown that non-timber products can add significantly to the value of the forests (Peters et al. 1989; Meulenhoff and Silitonga 1978). The value of less tangible goods provided by the forest, such as protection from soil erosion or habitat preservation, often goes unreported as well (Repetto et al. 1989). By not explicitly recognizing the importance of these goods and services, governments may allow the costly and often irreversible destruction of their most valuable assets. Indeed, a report by the World Resources Institute contends:

A country could exhaust its mineral resources, cut down its forests, erode its soils, pollute its aquifers, and hunt its wildlife and fisheries to extinction, but measured income would not be affected as these assets disappeared. (Repetto et al. 1989, 2)

Myers (1986) suggested that in western Amazonia, a square kilometer of humid tropical forest, if managed on a sustainable basis, could produce an overall annual income of at least $20,000 (i.e., $200/ hectare) from non-timber forest products. Although a specific timber harvest can have a much greater value than that of non-timber products, the net present value of non-timber production, when measured over a longer period of time, can exceed that of timber (Peters et al. 1989). Values of non-timber goods *in situ* have been estimated in a variety of tropical forest contexts. Although all of the estimates are not directly comparable, due to varied research and valuation methods, the potential opportunity cost of forgoing such a large stream of income is clearly high. In a brief literature review, Godoy and Lubowski (1991) note that the net present value of non-timber goods on a per hectare, annual basis has been estimated to range from $8 for wildlife alone in Sarawak, Malaysia (Caldecott 1987), to $3,327 for medicinal plants in Belize (Balick and Mendelsohn 1992; Godoy and Lubowski 1991). Depending on the range of species present, the interdependencies among species, and the production costs of sustainable harvesting, lesser-known non-timber goods have the potential to generate significant income and employment.

Some non-timber goods have already emerged from obscurity. Various commercial products presently cultivated on plantations, such as rubber, coffee, bananas, and cocoa, originated as non-timber forest goods. Other products, like rattan, spices, nuts, ivory, and dammar, which have been commercially traded for centuries, are still collected from the forest and in some cases have great current economic value. Semiprocessed and unprocessed rattan from Indonesia supports an export trade worth $100 million annually to that country in a typical year (Godoy and Rodrick 1990; de Beer and McDermott 1989), while worldwide the rattan furniture trade is estimated to be worth more than $2.5 billion/year (Siebert 1990). The Brazil nut harvest from Brazil has an annual production of 50,000 tons (Prescott-Allen 1982). In 1987, processed Brazil nuts sold for approximately $10/kg in the United States.

The collection and export of non-timber forest products can be a major source of foreign exchange in some countries where non-timber products have entered international trade and are taxed and recorded.

For instance, non-timber products contributed 36 percent of the total revenue of the Forest Department in India in 1980–81 (Tewari 1982). In Thailand, export revenues from non-timber goods increased by 81 percent from 1982 to 1987. Malaysia witnessed even more dramatic growth as non-timber export earnings grew by 480 percent from 1986 to 1988 (de Beer and McDermott 1989).

Beyond the commercial value of non-timber forest products, forests serve as a significant, and in many cases the only, source of livelihood for rural populations in developing countries (Meulenhoff and Silitonga 1978). Compared with other natural resources, forests not only can meet practically all the needs of rural populations, including the provision of materials for housing, food, medicinal products, condiments, and cord, but also represent a potential source of cash earnings.

Non-timber goods are often the basis for local craft production and small-scale industries. This can create employment both through self-employment, since the harvesting of these goods is labor intensive, and through secondary employment generated by the processing and marketing of these goods and their derivatives. Because the harvesting and collection of the products are often labor intensive, they can usually provide more employment than manufacturing. Employment in harvesting non-timber goods can also generate significant income for local gatherers and processors. Balick and Mendelsohn (1992) recently found that in Belize "bush masters"—gatherers of traditional medicinal plants—can earn from two to ten times the effective annual income of farmers on formerly forested land without threatening the long-term survival of the tropical forest. The study estimated the market value of known medicinal plants gathered in a 0.25 hectare area, including: negrito (*Simaruba glauca*), used traditionally by local Mayans for skin and digestive ailments; gumbolimbo (*Bursera simaruba*), a diuretic; China root (*Smilax lanceolata*), used for rheumatism and skin conditions; and cocomecca (*Dioscorea*), used for bladder infections and kidney "sluggishness." The study concluded that, after accounting for the costs of gathering, the forest had a net present value based on medicinal plants alone of $3,327/hectare, given a 50-year rotation—close to ten times the value of similar land cleared and used for intensive agriculture (Balick and Mendelsohn 1992).

A study conducted by the Environmental Defense Fund and Institute for Amazonian Studies (IEA) in Xapuri, Brazil (IEA 1989), indicated that local communities involved in the extraction of rubber and nuts earned an average annual income of $960 per family, which, combined with other nonmonetized activities such as fishing and hunting, resulted in an effective annual income of $1,500 per family. When compared with economically active wage earners in the general northern Brazil region, the rubber-tappers earned more than twice what was earned by half of the general population (IEA 1989).

In India, not only is family income augmented through commercialization of non-timber products but the income from certain non-timber products is higher than revenue from traditional sales of poles and firewood. In Pukuria village, in West Bengal, estimates of the income generated through gathering and processing of leaves from sal (*Shorea robusta*), cocoons, seeds, and gums surpassed take-home revenue from sales of poles and firewood by 500 to 700 percent (Malhotra and Poffenberger 1989). While returns on non-wood products averaged Rs (rupees) 7.6/hectares/day, returns on non-timber wood products averaged only Rs 1.4/hectares/day. Although local harvesters of poles and firewood may retain only 25 percent of sale proceeds (given government revenue requirements), even the gross marginal return on alternative products is higher than on wood products (Malhotra and Poffenberger 1989). Moreover, sal leaves, cocoons, and seeds are gathered sequentially through the year, whereas poles and firewood are gathered on a ten-year rotation. Collection of "minor" forest products in that village constituted the primary economic activity: village production totaled 10 metric tons of sal seeds annually, 700,000 leaf plates per month (four months/year) and, in December and January, over 300 wild tussore cocoons, each commanding four or five times the market price of cultured cocoons (Malhotra and Poffenberger 1989).

The case from Pukuria demonstrates that traditional harvesting of non-timber products in tropical forest regions has several other advantages. Under favorable year-round climatic conditions, and given a variety of non-timber goods, it is possible to spread harvesting and collecting operations through the year, or intersperse these activities with harvesting of field crops. Although fruiting is irregular for some species of tropical vegetation (see *Indigenous Food Products*), harvests of certain species can occur annually or semiannually, providing a steady flow of income to local workers. Economic failures are negligible: the variety of goods tends to diversify the financial risk associated with harvest of a single crop. Moreover, while large-scale timber harvesting is mechanized and requires large capital investments, gathering non-timber products is rarely mechanized and requires a relatively large amount of labor and modest initial investments. Therefore, production of non-timber goods is particularly suitable for countries with labor surpluses, a characterization that applies to many developing countries endowed with tropical forests and predisposed to trade in non-timber goods.

Competition between production of non-timber goods and other land uses is most likely to arise on lands endowed with fertile soil, which are most prone to agricultural conversion. While fast-growing trees, including the building phase and light hardwood species (see chapter 3), attain greater total volume under conditions of high nutrients and water access, many species can prosper on lands that would be considered marginal or unsustainable for traditional agriculture. Unstable surfaces, such as catchments and watersheds, therefore are most likely to demonstrate a comparative advantage for sustainable production of non-timber forest products, rather than traditional crops.

Non-timber goods might play a positive role in forest conservation and development. Their collection from the forests, if done on a non-destructive basis, may pose a negligible threat to the maintenance of a continuous forest and can result in minimal changes to the undisturbed primary forest. It is worth noting that no research appears to have evaluated the impact of harvesting non-timber goods, such as flowers and fruits, on the demography of plant species that produce them; similarly, the effect of hunting and fishing on the sustainment of animal populations is notoriously difficult to evaluate (Godoy and Wilkie 1991). Growing plant species that yield non-timber products may be incorporated in reforestation schedules for wasteland or logged-over forests. Some species are suitable as cover for hardwood timber trees, while many provide early yields, thus effectively reducing the discount rate on the investment in the timber crop. For instance, rattan, sugar palm (*Caryota*), and some medicinal vines and shade herbs greatly increase following the logging of wet-zone forests in Sri Lanka, producing early yields of importance to the rural economy (C. V. S. and I.A.U.N. Gunatilleke, personal communication). In the evergreen forests of the western Ghats, India, the Karnataka Forest Service employs forest-dwelling minorities to propagate tree species, planted in logged forests; the species yield not only timber but also dyes for the craft industries.

REASONS FOR THE NEGLECT OF NON-TIMBER FOREST PRODUCTS

Considering the potential and actual benefits that utilization of non-timber products of indigenous forest can provide, both for local populations and for whole countries, it may seem surprising that they have been so neglected by policy makers, managers, economists, and foresters. Most available explanations of this neglect can be classified in one or more of seven categories:

1. a lack of adequate information, especially economic and statistical data;
2. no established world markets, except for a few products;
3. an irregular supply of such products, and a lack of adequate quality standards;
4. a lack of processing and storage technology for many of the perishable products;
5. availability of artificial substitutes;
6. economies of scale achieved by plantations;
7. institutional bias toward large-scale investment in development.

The available literature on non-timber forest products rarely goes beyond descriptive studies of the products, their chemical contents, and traditional uses and is characterized by a lack of hard data on the economics and management of non-timber products (de Beer and McDermott 1989). Data are also insufficient on potential yield, on channels for distribution, and on economic value, especially of those products that are not yet commercially exploited but have served as consumption goods for rural populations.

Non-timber forest products are often associated with traditional uses that are not widely known (Clay 1988). They are seldom marketed through known channels that would add to a nation's gross national product (GNP). Timber markets are already developed and well integrated, while markets for most non-timber goods remain undeveloped. In the case of timber, numerous species of trees are marketed through the same channels under a common label; due to their variety, those non-timber products suitable for domestic or international trade must be marketed individually. Thus, market-oriented development officials seldom give much thought to these products when considering management options for forested land. Foresters, too, have tended to ignore these products, having been largely schooled to think exclusively in terms of standing volumes and growth rates of timber species. Available economic and statistical data are limited to those few products that are either exported or have a known market value, such as rattan, Brazil nuts, and latexes.

Non-timber goods also tend to be overlooked because, as individual products, they do not contribute significantly to export earnings. Especially when expressed as a percentage of timber exports, the value of individual non-timber products is relatively small. For example, 80 to 90 percent of the world production of rattan, usually considered the most economically successful non-timber product, is produced in Indonesia (Manokaran and Wong 1983, 299). In 1975 this represented only 4 percent of the value of Indonesia's entire timber exports (see table 5.1). Nevertheless, a decade later, by 1986, this value had increased to approximately 10 percent of the timber export value (R. Godoy, 1988, personal communication). Although the demand for rattan has increased since 1962, its value, expressed in terms of the hardwood timber trade, therefore remains small.

These low relative values can be misleading. In absolute terms, the added value that rattan contributes to the forest-based economy remains significant, since Indonesia is also the world's largest exporter of tropical hardwood timber: when tallied collectively, non-timber products can markedly improve a country's trade balance. The values of non-timber goods should be assessed by comparing the production and market of both timber and non-timber goods (and services) on the same area of land. If all non-timber goods in a hectare of forest are considered, the value added to a forest can be substantial and might even exceed that of timber in some cases (Peters et al. 1989).

The unpredictable yield of certain non-timber products constitutes a serious drawback for commercial interests and discourages economists from considering non-timber production more seriously. This is particularly true in the case of fruits, nuts, and wild game. The lack of sufficient volume per hectare of certain species can prevent the development of a regular supply (Meulenhoff and Silitonga 1978). Additionally, the distribution of many species producing non-timber end

Table 5.1

EXPORTS OF NON-TIMBER AND TIMBER PRODUCTS IN INDONESIA, 1973–1983
(IN MILLIONS OF U.S. DOLLARS)

Year	Value of Non-Timber Exports	Value of Timber Exports	Non-Timber Exports as Percentage of Timber Exports
1973	17.0	583.4	2.9
1974	24.9	725.5	3.4
1975	21.6	527.0	4.1
1976	34.7	885.0	3.9
1977	48.3	943.2	5.1
1978	58.6	1,130.6	5.2
1979	114.0	2,172.3	5.2
1980	125.6	1,672.1	7.5
1981	106.0	951.8	11.2
1982	120.0	899.4	13.3
1983	127.0	1,161.1	11.0

BASED ON: Indonesian Forest Department Statistics in Repetto and Gillis 1988.

products can be patchy through the forest. Therefore, the development of markets for such products is difficult due to irregular supply and problems in maintaining quality standards. Many products, especially forest fruits and vegetables that are sensitive to moisture and temperature, are highly perishable and cannot be kept fresh for a long time. They can be commercially exploited only if adequate technology is developed for their processing, transportation, and storage. This is an area where further research is required.

Alternative production systems—both synthetics and monoculture plantations—of highly valued non-timber goods also contribute to the low valuation of these goods *in situ*. Large chemical industries in developed countries have often successfully produced substitutes for chemically active plant substances originating from tropical forest species. In the food industry, the use of substitutes is extensive, and includes synthetic substitutes for spices (e.g., vanilla, which can now be produced in tissue culture, and fruit essences), food dyes, and preservatives. The same applies to medicinal and pharmaceutical plants. Natural rubber (*Hevea brasiliensis*) has been partially replaced by neoprene and other synthetics. Rubber has also been produced through intensive management of monoculture plantations, which led to a substantial drop in exploitation of natural rubber (rubber nonetheless remains a significant foreign-exchange earner even for Brazil) (Macmillan 1991, 449–51). Harvesting medicinal and cosmetic plants from natural forests faces the same type of competition.

Another reason for neglect of non-timber goods is institutional. Policy makers and lending agencies have favored large-scale development projects, and therefore have seldom recognized the potential returns from non-timber forest goods. Although profit opportunities from capital investment by state or commercial investors are relatively scarce, social and economic returns from development of non-timber production can be significant for the rural poor.

Classification of Non-Timber Goods

Because of the variety of non-timber products, more than one classification scheme is possible. Systems can be based on one or more of the following criteria:

1. the group of organisms from which the products originate;
2. the specific parts of plants or animals yielding valued products (roots, trunk, bark, leaves, fruits, seeds, pelts, bones, or horns);

3. their end-uses in trade and industry;
4. the manner in which they are collected or harvested; or
5. their places of origin in the forest.

To serve industry, for example, a classification system could be based on chemical properties, e.g., latexes, carbohydrate sources (starch, fruits, bamboo shoots), oils and fats, or gums and resins.

For this study, non-timber products are classified according to their importance and potential as income earners when harvested from the forest. This approach yields four broad categories:

1. raw industrial inputs;
2. indigenous species used for food, including nuts, fruits, locally consumed vegetables, and wildlife;
3. products highly valued, either locally or in foreign markets, for reasons of rarity or cultural significance; and
4. energy products.

Classification according to income generation also yields important distinctions concerning potential industrial growth and opportunities for increasing value added prior to export. Raw materials for craft-based industries such as rattan, bamboo, and skins represent non-timber products with significant potential for local industrial development and significant increases in value added. Likewise, forest-based food products, including nuts, some fruits, and bushmeat, are goods with emerging international markets as well as strong domestic markets and opportunities for creating local income.

Other products are capable of significant return not for their industrial use or consumption as food but because their very rarity or traditional cultural importance renders them highly valuable, in both domestic and international markets. Items considered to have special attributes, such as the rhinoceros horn, or particular local applications, such as spiralwoods used for coffins and home shrines, or medicinal plants, can yield high returns for gatherers and producers. Although medicinal plants have been highly publicized in Western industrialized countries as potential foreign-exchange earners, plants yielding specialized chemical products are at risk of replacement through development of synthetics or tissue culture, once their value approaches a certain threshold. They nevertheless remain the medicines of choice in many developed and developing nations.

Energy products are deemed to have limited potential for harvesting from mixed-hardwood-production forests and plantations. Research on tree cultivation for charcoal indicates that profits can be realized from carefully managed plantations (Murray 1988: see below). In principle, mixed-species management of forests could include charcoal production for domestic consumption; the regulation required to ensure protection of juvenile timber species is difficult to enforce. Fuelwood, because of its higher weight per unit of energy and the need to cultivate pioneer species, is least compatible with mixed-hardwood cultivation.

Raw Materials for Industry

Non-timber products valued for their industrial use range from harvested vegetable material to animal products. Rattan and bamboo are the two most economically valuable non-timber forest goods that fall into this category. Rattan cultivation has proved profitable both for smallholders and the national economy in southern Borneo, Indonesia, generating an economic internal rate of return of 21 percent for green canes and 22 percent for processed canes (Godoy and Tan 1989, 353). Rattan's economic and financial viability as a non-timber product of the tropical forest is well documented (see appendix 5.1). Other economically promising forest plant species that provide raw materials for craft-based industry include bamboo, kapok (once a major export and still used for mattresses in Africa), Borassus palm (Senegal), and Raphia palm, widely used in Africa for wickerwork and manufacture of cord and rope.

Oil palm and cocoa are industrial crops requiring highly intensive management. Therefore, they cannot be grown in mixture with lower-value tree crops such as timber, and are usually produced in plantations. The only exceptions are the palms, such as *Orbignya martiana* (babassu palm) and *Jessenia bataua* (see table 5.2), which grow in very high densities in some natural forests in South America (Anderson et al. 1991). Due to their number and the areas they cover, these palms can potentially become a major source of oil production. For example, babassu palm forms a dominant cover over millions of hectares in central and northern Brazil. These palms demand full sun, however, and are therefore in direct competition for space with timber trees.

Wild game, including mammals, fowl, reptiles, and fish can also be considered non-timber forest products that have potentially high commercial value, but are not generally included in calculations of forest values. Worldwide, animal skins appear to have the greatest potential as raw materials for craft-based industry. The final processing of furs and pelts (where most added value is gained) takes place in the importing countries, but the possibility of local industrial development reliant on

Table 5.2
MOST ECONOMICALLY VALUABLE FRUIT SPECIES OF AMAZONIA

Species	Use	Distribution	Density (adults/ha)	Yield (metric tons/ha)	Value ($US)
Myrciaria dubia	fruit	NW and W Amazonia, along banks of ox-bow lakes	1,124	11.1	6,661
Grias peruviana *Grias neuberthii*	fruit and oil	Seasonally flooded forests in W. Amazonia	192	2.3	4,242
Mauritia flexuosa	fruit	Flooded swamp forests throughout Amazonia	223	6.1	1,525
Jessenia bataua	fruit and oil	Upland forests throughout Amazonia	104	3.5	306
Orbignya phalerata	fruit, oil, and charcoal	Upland forests in central and northern Brazil	223	1.5	23

BASED ON: Peters et al., 1989. Reprinted by permission of *Nature*, vol. 339, pp. 655–656. Copyright © 1989 Macmillan Magazines Ltd.

forest management exists. In all cases, though, creating conditions favoring sustainable rates of harvesting has proven to be difficult, and may be impossible; once again, further research is required.

Indigenous Food Products

Fruits, nuts, and bushmeat collected in the forests have traditionally been consumed by rural populations. However, in a few exemplary cases, they have also generated significant foreign exchange and local employment. Brazil nuts (*Bertholletia excelsa*) yielded $40 million in foreign exchange in 1988 for Brazil (CGBD 1989). They were first imported to the United States in 1810. By 1982, Brazil nut imports had grown to 23 million pounds and were sold at retail for $1.30 to $1.50 per pound, a price that continued to rise through the 1980s. Fueled in part by the successful marketing of Rainforest Crunch ice cream and other products, the growth of Brazil nut exports is not without complications. Profit sharing with local harvesters, quality control, transportation challenges, and supply limitations constitute major sources of concern for both marketers and local government agencies. In addition, although Brazil is the major producer of Brazil nuts, with an average annual output of over 50,000 tons, the destruction of many of the Brazil nut stands in the conversion of forests into farmland has cut into commercial production (USDA 1976).

In Borneo, the commercial illipe nut (known more correctly as *tengkawang* and unrelated to the Indian illipe) is a major forest product. However, it is cyclical in production, with major fruiting years occurring irregularly at approximately four-year intervals (Ashton et al. 1988). Less than 10 tons are exported from Sarawak in 35 percent of the years, but up to 28,000 tons are exported in others (Caldecott 1987, 83). Illipe production directly affects two other non-timber goods: bearded pigs (*Sus barbatus*), which seek illipe fruit as a preferred source of food, and rattan. During illipe harvesting season, rattan collection is drastically reduced because the workers shift to the more profitable collection of the nut.

Despite these problems, Brazil nuts and illipe nuts represent two examples of a growing cadre of forest-grown food products that have been marketed internationally from forests with success. Former forest fruits such as bananas, mangoes, macadamia, and cocoa are now grown commercially in plantations for the world market, while many others are planted for local and regional markets. Several species such as tamarillo (*Cyphomandra crassicaulis*) and Feijoa (*Feijoa sellowiana*) have potential for further market and plantation development, while there are undoubtedly many more species in the forest that warrant attention as potentially valuable products never before marketed beyond the village. The forest is generally valued therefore as the original source of new major commercial products, but a constant and maximal supply at competitive prices cannot be sustained when commercial demand increases.

Before the large-scale exploitation of these products can begin, certain basic questions must be answered. The relative abundance of the species to be harvested and its role in the food chain are crucial factors in determining the potential costs and benefits of commercial exploitation. For example, the opportunity cost of large-scale harvesting of certain fruits and nuts may include threatening the survival of a local herbivore, the value of which may be significant in economic as well as ecological terms.

Little has been written about the density, phenology, and productivity of fruit resources growing in nature. One exception is the study of fruit trees in Iquitos, Peru, by Peters et al. (1989). (The methods used in this study are elaborated in chapter 7, appendix 7.2.) In their study, Peters et al. singled out three fruit species—*Myrciaria dubia*, *Grias peruviana*, and *Mauritia flexuosa*—as the most valuable in the Iquitos region (see table 5.2). All three trees grow in high densities in the seasonally flooded forests of the Peruvian Amazon. Their fruit is used both for local consumption and sale (see table 5.2).

Myrciaria dubia, also known as camu-camu, is exceptionally rich in vitamin C, its fruit containing 2,000 to 3,000 mg of ascorbic acid per 100 grams of pulp (Ferreyra 1959; Roca 1965). The fruit is used in juice drinks, ice cream, and the popular Peruvian liquor, *camu-camuchada*. The species is especially abundant in the Peruvian Amazon, where it forms large, monospecific stands in riverine forests with densities of up to 8,500 stems/hectare (Peters et al. 1989). If the entire fruit production were collected and sold in the market, the authors estimate that gross revenues could total $6,700/hectare annually.

The fruits of *Grias peruviana* are collected and consumed in many areas of the Peruvian Amazon, both for eating and for oil extracts. The species forms dense understory populations that are extremely shade tolerant. The tree is also occasionally cultivated in home gardens.

Of the tree species that Peters et al. examined, *M. flexuosa* is the most versatile. Almost every part is used by the local population: fruits are eaten raw or processed, tea and bark are used medicinally, and the wood provides relatively good timber for light construction.

In terms of total yield, the natural populations of *M. dubia*, *G. peruviana*, and *M. flexuosa* examined in the study compared favorably with intensively managed populations of fruit trees. For example, avocado populations annually produce from 2 to 10 metric tons of fruit/hectare (Ochse et al. 1961), while mangoes usually produce from 3.5 to 6.5 metric tons of fruit/hectare annually under cultivation (Hayes 1953). The productivity of the fruit trees examined (table 5.2) ranged from 2.3 to 11.1 tons/hectare, a remarkable yield considering that neither fertilizers nor silvicultural practices are employed.

The forests examined by Peters et al. (1989) are atypical of the Amazon valley as a whole. The figures derive either from hill forest on soils sufficiently fertile to be arable or from riverine forest where, as in the case of freshwater fish, the resource is receiving nutrient input from the floodwaters of an entire catchment. Results cannot be extrapolated to dryland forests, which form the majority. Furthermore, the figures cited in Peters et al. represent potential, not actual, yields, and assume no foraging by wildlife.

Although Peters's calculations were based on extreme assumptions, namely that all fruit produced by each population is collected and that the fruit prices remain stable regardless of the supply, they do provide an intriguing example of the economic potential of some forest resources. Increasing the current level of exploitation of *M. dubia*, *G. peruviana*, and *M. flexuosa* might therefore yield economic benefits with only a minimal ecological impact on the forest. In fact, Peters et al. may have provided us with an example of successful competition by non-timber goods with timber (see appendix 7.2). In such a case, non-timber products might be given silvicultural preference in the stand they studied. Experience suggests, however, that this is a rare circumstance, and most likely to occur on land that will eventually be profitably converted to agriculture (including fruit-tree plantations).

Substantial local markets exist for mushrooms and the bulkier forest vegetables, notably palm hearts and bamboo shoots, but export markets remain relatively limited. Palm hearts, derived from a number of palm genera, are currently harvested from Andean forests, substantially for export. But the removal of the heart, which is the terminal bud, generally kills the plant, and the sustainable management of palm populations in natural forests for this purpose has yet to be demonstrated. Given that the number of bamboo plantations in Thailand has been increasing steadily, it appears possible that monoculture bamboo plantations are more profitable than the collection of bamboo shoots from the forest.

As Prance (1984) suggests in his study of the use of edible mushrooms by Amazonian Indians, there is potential for cultivating mushrooms as part of forest management. The process of swidden agriculture, for example, creates an ideal situation for the growth of fungi on the many logs which are left lying in the fields and which rot in the understory of the woody fallow. Many of the fungi, such as *Favolus brasiliensis* and *Polyporus tricholoma* are characteristically found on rotting logs. Because they flourish under the combination of humid climate, an abundance of fallen logs, and standing tree trunks, they might be produced in logged-over forests, since damaged trees continue to die and fall long after the logging crews have left.

Okafor (1979) discussed the economic importance of West African food tree crops such as *Elaeis quineensis*, *Irvingia gabonensis*, *Treculia africana*, *Pentaclethra macrophylla*, *Cola acuminata*, and *Dacryodes edulis*, and also *Spondias mombin* (indigenous to South America and introduced to Africa). Apart from being used as sources of fresh fruit, the leaves, stems, bark, fruits, seeds, flowers, and tubers of many of these edible woody plants are used locally and regionally in the production of seed oil and fat, fruit drink, wine, fruit jams and jellies, animal feed, chewing sticks, stakes, structural materials, and mulch. Many of the edible tree crops have very high concentrations of protein (12 to 48 percent), fats (10 to 72 percent), and such minerals as calcium (1 to 2 percent) and iron (50 to 450 ppm) (Okafor 1979). Available forest enumeration and inventory data demonstrated that some edible forest trees are widely distributed throughout the forest zone of Nigeria, although their density per hectare is low (less than 2.5 trees/hectare above 61 cm diameter at breast height [dbh]). With greater protection and some degree of cultivation, the density of edible trees has been increased significantly. The gross revenue per tree per day for palm wine ranges from $0.26 to $1.99, and the revenue per day to each forest farmer may be appreciable, up to $17 depending on amount of work and number of trees tapped.

Other important food sources from forested regions include bush-meat and fish. In South America, according to Dourojeanni (1985, 421), significant wildlife species include tapir (*Tapirus terrestis*), peccary (*Tayassu pecari*), deer (*Mazama americana*), and the collared peccary (*Tayassu tajacu*). For Africa, de Vos (1977), lists bushback (*Tragelaphus scriptus*), grey duiker (*Sylvicapra grimmia*), and giant rat (*Cricetomys gambianus*), while species in Asia include rusa (*Cervus unicolor*), kijang (*Muntiacus spp.*), pig (*Sus barbatus*) (see appendix 5.2), and various species of monkeys.

Among all wildlife species, however, it is fish that yield the greatest return in foreign exchange: "Fisheries dominate trade in wild game. In the five-year period, 1974–78, the average annual value of developing country fishery exports was $3.2 billion (FAO 1980), second only to timber at $4.5 billion" (Prescott-Allen 1982, 26). While the data refer principally to saltwater fisheries, it is worth noting that they, too, rely on forests for survival; coastal mangrove forests provide crucial spawning grounds for many species.

Wildlife is also critical to regional and local economies. Locally, wild animals and fish often are the only source of nonvegetable protein for the rural population (Clay 1988). Wildlife is most important in the savannas and woodlands of the tropics, where the grass field layer and more nutritious leaves of certain deciduous woody species provide considerably more browse and forage than in evergreen forests. Wildlife accounts for between 20 and 90 percent of total animal protein consumed in the Benin Republic, Cameroon, Ghana, the Ivory Coast, Liberia, and Nigeria (Ajayi 1979). In Botswana, 60 percent of the animal protein consumed annually is derived from wildlife (FAO 1969). These animals range from elephants to small ungulates as well as rodents, reptiles, and birds. Wild animals yield more animal protein in the form of lean meat per unit live-weight than do domesticated animals under the same conditions (see table 5.3). Afolayan (1980) estimated the total value of bushmeat in Nigeria at $30 million and the total value of naturally produced protein food at $100 million. During 1965–66, a total of 617,000 tons of bushmeat worth $20.4 million was consumed, compared with 714,000 tons of beef worth $27.8 million in southern Nigeria (de Vos 1977). The annual bushmeat consumption in Ghana was estimated at 8,486 tons valued at $7,358,172 (de Vos 1977).

In South America, too, wildlife serves as an important source of

Table 5.3

TERRESTRIAL WILDLIFE AS A SOURCE OF FOOD IN SELECTED COUNTRIES OF AFRICA

Country	Food Consumption and Species Concerned
Botswana	Estimated consumption of game meat is 60% per person annually. Over 50 species of wild animals, ranging from elephant to rodents, bats, and small birds provide 90.7 kg of protein per person annually.
Ghana	About 75% of the population depends on traditional sources of protein supply, mainly wildlife including fish, insects, caterpillars, and snails.
Ivory Coast	In the northern part of the country, 27 grams of bushmeat were consumed per person per day.
Nigeria	Game constitutes about 20% of the mean annual consumption of animal protein by people in rural areas.
Zaire	75% of animal protein comes from wild sources.

BASED ON: de Vos 1977.

protein. The floodplains of the Amazon Basin of Peru and Brazil harbor freshwater fisheries that rely on forested watersheds for nutrients and detrital food (see appendix 5.2). In the Peruvian part of the region, fish from the Amazon River are the major source of protein (see table 5.4). In summary, plant and animal species from the tropical forest can generate significant income and employment for hunters, gatherers, and traders.

Wildlife provides a relatively inexpensive source of animal protein for rural populations. Cultivation of these species in the context of multiple-use management of tropical forests should incur low management costs when appropriate forest policies are followed. Development of policies that ensure sustainable production under conditions of increasing population and demand poses the most formidable challenge.

Non-Timber Products Valued for Reasons of Tradition or Uniqueness

Many non-timber forest products derive their value neither from industrial applications nor from direct consumption as food, but because they are viewed as culturally significant or unique. These products include not only plants with special chemical properties but animal and insect species valued for their rarity and commodities locally produced and important for reasons of culture and tradition.

Latexes, resins, cosmetics, condiments, and other products that are exploited for their unique chemical properties may offer high value-added potential when collected and semiprocessed in the wild; however, once their value in international trade passes a certain threshold,

Table 5.4

MEAT CONSUMPTION FROM WILD ANIMALS AND OTHER ANIMAL
PROTEIN SOURCES IN THE AMAZON REGION OF PERU
(IN GRAMS OF MEAT DAILY PER CAPITA)

Sources	From Pucallpa to Nauta (Ucayali River)	In Jenaro Herrera (Ucayali River)
Fish	135.6	158.3
Game	52.0	75.8
Farmyard birds	22.1	25.7
Pigs	12.0	10.2
TOTAL	221.7	270.0

Reprinted with permission from Marc Dourojeanni, "Over-Exploited and Under-Used Animals in the Amazon Region," in G. Prance and T. Lovejoy, *Amazonia.* Copyright 1985, Pergamon Press PLC.

their production in developed countries is likely to be replaced by synthesis or biotechnological processes that mitigate supply uncertainties; strong traditional preference for the natural product in some regional and local markets will tend to preserve the income-generating ability of these products. Demand for medicinal products remains strong and stable in many countries, where up to 80 percent of the population may continue to depend on traditional remedies (WHO 1977, 46). This can be due to poverty in situations where imported medicines are too expensive or to retention and accretion of their value to prospering civilizations. As one example, African forest plants produce pharmaceutical gums (*Acacia senegal*, *Periserianthes* (*Albizia*) *falcataria*, *Zanthosoma caprestris*), disintegrants (*Manihot esculenta*, *Zanthosoma sagittifolium*), sweetening agents (*Cola caricifolia*, *Napoleona imperialis*, *Triclisia subcordata*), and lubricants (*Irvingia gabonesis* var. *excelsa*) (Oboriah 1986). In some species (*Thaumatococcus daniellii* and *Dioscoreophyllum cumminii*) the sweetening principal is protein-based and 3,000 to 5,000 times as sweet as sucrose on molar basis (Kio 1987). In these cases, the major challenge is to create conditions favoring sustainable use as populations and demand increase. It will remain economically profitable at least to grow resin and condiment-producing plants in mixed-species home gardens in many cases, while the social justification for growing medicinal plants likewise will persist for the foreseeable future.

Other examples of forest products valued for reasons of tradition or uniqueness include scented woods. Abnormal tissues in the wood of gaharu (*Aquilaria malaccensis*) trees in tropical Asia have continued to grow in value as an incense in the Arab world, as well as an export to industrialized nations, where it is used as a scent in perfume manufacture. Sandalwood, also, is a significant income earner for its unique scent and is produced in many tropical countries for local as well as foreign consumption.

Rare animal species can provide both significant foreign exchange and local income. Butterflies, for example, generated close to $100 million in exports worldwide by the eighties (Collins and Morris 1985, cited in Aruga and Saulei 1990). Major exporting countries include Malaysia, the Philippines, India, and Taiwan, but other suppliers are developing in Africa and Central America (Sayer 1990). Butterflies in Papua New Guinea have generated significant revenue for local farmers. In the Morobe province of Papua New Guinea, over 500 farmers are engaged in butterfly ranching: prices on the world market range from $1 to $20 depending on their rarity. Revenues generated through the Insect Farming and Trade Agency, which has control over all insect exports from the region, totaled $180,000 for the first three years of the agency's existence (1978–81) (Aruga and Saulei 1990).

Animal products may also be valued locally and regionally for cultural reasons. Like many other bird species in Papua New Guinea, the cassowary has high value not only as protein for the diet of Highlands residents but also as part of a brideprice, as a form of traditional compensation payment and as a source of decoration (Aruga and Saulei 1990). With the advent of a cash economy in many rural areas, commodities like the cassowary can be valued in terms of their cash equivalents: alive, this bird is worth $1,125 (Aruga and Saulei 1990, 43).

Energy Products

Production of non-timber goods for energy use represents an almost exclusively domestic industry. The bulk of fuelwood and charcoal is consumed locally and regionally by indigenous populations. However, although not a major producer of foreign exchange, energy products from tropical forests can be a major source of income for local producers.

Reliance on charcoal as a source of fuel appears to be increasing. Fuelwood and charcoal production worldwide grew 28 percent between 1975–77 and 1985–87 (WRI 1990). In South America, fuel and charcoal accounted for more than two-thirds of all industrial roundwood production in 1985–87; in Asia the figure was close to 75 percent, and in Africa, 88 percent of roundwood was produced for energy purposes (WRI 1991). Although such figures do not distinguish between natural forests and plantations as sources of energy supply—an important distinction in need of further research—the growth in fuel production provides ample evidence that energy is an increasingly significant output of tropical forests generally.

Charcoal is usually produced from trees whose wood has high calorific value, that is, low water and high carbon content. Good charcoal producers are partially confined to families such as *Fagaceae, Leguminosae, Myrtaceae*, and *Rhizophoraceae*. Many wood species appropriate for charcoal production are also quality timber producers.

Unlike fuelwood, charcoal has a value per unit of weight and volume sufficient to justify regional trade, provided that transporting the raw wood to the kiln is restricted to short distances. Charcoal is also preferred because of its steady and concentrated heat; it supplies twice the amount of heat per unit weight of wood (Myers 1984). Consequently, charcoal presents two possible sources of additional return in the management of natural hardwood timber forests. The first is in cost recovery, and perhaps profit generation, from thinning undertaken primarily for timber regeneration in silviculture operations (Wyatt-Smith 1963). Here, stringent supervision is essential to ensure that individual trees marked for eventual timber production are not culled by the charcoal harvester. The second is in post-felling cleaning operations. This can be carried out with minimum supervision in uniform felling systems, including clear-strip fellings (see chapter 3).

Currently, only a few studies on the marketable value of charcoal have been conducted. Murray (1988) estimates that with current market conditions in Haiti, charcoal harvests could yield gross revenues of $1.50 per tree. He calculates that a hectare of land can hold 2,500 trees, which can be harvested in four-year intervals. This translates to potential gross revenues of $3,750/hectare over each four-year period. This figure does not include potential revenues from other crops grown on the plots.

Trees that produce fuelwood are generally limited to those whose wood has relatively high calorific value and incendiary qualities. Fuelwood, like charcoal, can be derived in managed hardwood forests in the form of thinnings from silvicultural operations prior to harvesting, or cleanings at the time of harvesting. Unlike charcoal, fuelwood cannot profitably be both gathered in small volumes per unit area and transported long distances for sale, owing to its low unit value. Consequently, profitability is heavily dependent on access to local markets, while meeting the silvicultural harvesting requirements is unrealistically dependent on product profitability and strict supervision to ensure that regeneration of desired hardwood species is not reduced (see chapter 3; for a historical account, see Wyatt-Smith 1963). Fuelwood harvesting, therefore, rarely represents a means to add commercial value to managed natural hardwood timber production forests, though deadwood and litter have widespread social value to forest-fringe communities.

In the case of multiple-species plantations, fuelwood is most profitably produced by pioneer species that coppice on short harvesting rotations. This too is incompatible with mixed-species plantations where a longer-lived, taller, less frequently harvested shelter for the slower-growing hardwood species is required (see chapter 3).

REGIONAL ASSETS AND OPPORTUNITIES

All non-timber goods share two universal characteristics. First, the sustainability of species with obvious economic value is subject to constraints. If their economic utility is overlooked and ignored, they face the conversion of the habitats on which they depend; if, on the other hand, their economic value is evident, they are likely to be overexploited, often to the point of economic extinction because secure ownership rights are difficult to establish (de Beer and McDermott 1989). Second, non-timber goods provide food, shelter, and employment for millions of rural people whose livelihood would be in danger without them. These characteristics are manifest differently from region to region.

Asia

Because of the nature of the forest canopy in many parts of Asia, species producing non-timber goods appear to be less abundant compared to their counterparts in other forested regions of the world. Many Asian forests are dominated by timber-producing canopy trees of the family Dipterocarpaceae, whose presence may cause competitive exclusion (shading) of species yielding other goods (Wyatt-Smith 1963). Therefore, fewer understory non-timber-producing trees per unit of area are likely to be found in lowland equatorial Asia, in comparison to South America or Africa. The main non-timber products are rattan, bamboo, bearded pigs, illipe nuts, various spices (e.g., cardamom), various fragrances (e.g., gaharu, eaglewood), resins (e.g., dammar), wild fruits, wild sage (*Metroxylon* and *Eugeissona* palms), palm sugar, honey, live animals, and skins. Philippine and Australian forests yield valuable nut species (*Canarium ovatum* and *Macadamia*, respectively) and nutmeg (*Myristica fragrans*) originates from the Moluccas, but few comestible fruits occur east of Sulawesi, perhaps because there are no indigenous primates.

Rattans are the second most valuable forest products (after wood) in Southeast Asia (Menon 1980). Indonesia supplies 90 percent of the world rattan demand and has revenues of approximately $100 million/year. The trade of finished rattan goods is worth nearly $1.2 billion (see appendix 5.1).

World bamboo production is currently more than 10 million tons a year, originating primarily in Asia (Sharma 1980). Bamboos are widely cultivated but are also extensively collected from the wild. The most important industrial use of bamboo is in paper manufacture: in India, an estimated 2 million tons (dry weight) of bamboo a year provide 600,000 tons of paper pulp (Lessard and Chouinard 1980). The primary importance of bamboo, however, is for diverse uses in the village economy, including construction, water ducts, flooring and matting, and crafts.

Asia is the region that exports the largest number of live primates and reptile skins: Indonesia and Thailand together produced 4.7

million skins in 1986—close to half of the world's exports (WRI 1990). Fish and wildlife are critical to the local population as well: in Thailand more than 60 percent of the rural people depend on fish as their main source of animal protein (Brennan 1981, 24).

There appear to be major differences between Southeast Asia and Central America regarding fruit trees. Central American trees are much more regular in fruiting and they often exist in dense stands. In Asia, fruit trees tend to be more dispersed, and those in equatorial regions exhibit irregularity in flowering. It has been suggested that the lower density of arboreal mammals noted in Asia, relative to other forested regions, is due primarily to limited availability of high-quality plant food (Janzen 1974). The lower food availability may not only be due to competition between fruit trees and the dense, full emergent canopy of Dipterocarpaceae; the explanation may as likely lie in frequently low soil fertility and associated levels of productivity (Janzen 1974).

These characteristics of Asian fruit trees do not suggest that Asian tropical forests necessarily offer less promise as sources of non-timber food crops than Central American forests. While the concentration of trees in Asia is lower, the diversity and number of species is at least as high. Moreover, the potential for growing fruit trees and vines in mixed plantations with hardwood species is particularly promising in Asia in view of the highly developed traditions of mixed-species home gardens, and hence the existing indigenous experience and sociopolitical acceptability. Yet there is much room for improvement in plantation design, species choice, genetic variation, quality control, storage technology, and market development.

Tropical America

Of the three continental regions of the humid tropics, Central and South America differ fundamentally from the other two in the low total economic returns gained from their forests. For example, while returns from Sarawak, East Malaysia, one of the least populated regions of Asia, indicate total export values of timber goods in 1980 at $11,632 per annum per square kilometer of forest (Sarawak Government Statistics), the states of Rondonia, Acre, and Amazonas averaged $42.50 per annum gross per square kilometer for total production in all markets of all goods (based on Allegretti and Schwartzman 1986). If Caldecott's (1987) upper estimate of the value of forest game in Sarawak is accepted, then the total value of non-timber goods in Sarawak minimally approximates 7 percent of timber exports, or $800 per square kilometer, while the value of the non-timber goods for the Brazilian states for

1980 was a meager $30 per square kilometer, albeit a 187 percent increase over the previous decade. These regional statistics place the study by Peters et al. in a broader perspective. Their estimated gross profit of $6,700 per hectare in the Iquitos region of the Peruvian Amazon would seem to bear little relationship to larger regional realities. On the other hand, the current rate of exploitation of Sarawak's forests is patently not sustainable (Anon. 1990).

Significant inherent biological differences exist between the forests of Asia and Tropical America, notably in timber stocking (especially the absence of timber-producing Dipterocarpaceae outside Asia) and in flowering and fruiting phenology. While South America currently lacks the institutional, marketing, and transportation infrastructure to respond easily to increasing demand for timber in world markets, it is these biological differences between the two regions that account for the different roles played by forests in the international economy. In the absence of a family of dominant fast-growing hardwoods such as the Dipterocarpaceae in Asia, it seems that most lowland South American forests lack the biological resources, overall, to yield timber revenues comparable to those generated in the Far East.

Latin American non-timber products are distinguished by their variety. Wildlife and fish are as significant economically in Latin America as rattan is in Asia. Fish and wildmeat (peccary, deer, tapir) account for 85 percent of the meat consumed in the Amazon, despite the availability of cattle and pigs (Pierret and Dourojeanni 1966). One reason for this is the relatively high price of meat from domesticated animals (see appendix 5.2).

Other specialized products, particularly latexes and Brazil nuts, are also important regional goods. Despite the development of synthetic substitutes, natural rubber remains the most important of the latex products. More than 20,000 tons of wild rubber (*Hevea*) was harvested in Brazil in 1979, with a total value of $20 million (Peters et al. 1989); this is an admittedly insignificant sum compared to natural rubber revenues in Asia, where rubber was introduced early in the twentieth century and has been successfully cultivated ever since. Brazil nuts and other products are often collected in the same areas where rubber tapping takes place. In fact, the rubber-tappers often extract 14 to 15 native products. In combination, these goods can generate significant income. For example, the Amazonian state of Acre recorded earnings of $26.2 million in 1980 from rubber, Brazil nuts and other fruits, and wild meat (Allegretti and Schwartzman 1986). A total of 70,000 rubber-tappers make their living by gathering latex and Brazil nuts in the Brazilian Amazon. However, because of the open access to the resource and inadequate competition in the marketing of rubber and other non-timber products, resource rents are partly dissipated by excessive extraction and partly absorbed by monopsonistic traders: precise estimates are lacking and require further research. In this regard, the establishment of extractive reserves is a necessary, but not sufficient, condition for enhancing the value of non-timber products and ensuring that a greater portion of the rents remains with collectors.

Africa

Africa's main non-timber product is wildlife. While many countries derive considerable income from game meat and meat by-products (see *Indigenous Food Products*), living animals also generate income. Kenya's tourism trade directly generates an estimated $350 million to $450 million annually (Dixon and Sherman 1990, 184), with much more income accruing to Kenyans in the form of associated industry and spending through the multiplier effect (see *Tourism* in chapter 6 for a discussion of tourism in other tropical forest regions).

While the most value added in African game occurs through tourism, the industry is focused on the dry savanna forest, where wildlife is both plentiful and visible. The moist tropical forests of Africa generate foreign exchange primarily through export of live animals. Ghana, for example, exported sixteen species of wild animals (see table 5.5), earning $0.5 million in 1985. Tanzania exported more than 84,000 parrots in 1986 (WRI 1990); while the actual value of the Tanzanian trade is not known, assuming an average world price equal to that attained by Ghana for the African Gray Parrot ($30/head—parrots often command far higher retail sums in developed-country markets), total exports of parrots alone would have generated over $2.5 million.

Game is not normally competitive with timber production through natural forest management in the humid tropics. Only in the forest-

Table 5.5
LIVE WILD ANIMAL EXPORTS FROM GHANA, 1985

Species	Number	Value ($US)
African gray parrot	9,580	287,400
Royal python	4,449	35,592
Scorpion	2,170	3,038
Agama lizard	1,600	2,240
Green parrot	400	6,000
Rock python	254	2,032

BASED ON: Ntiamoa-Baidu 1987.

savanna margin is trampling and browsing a major threat to hardwood regeneration, where herds of large mammals may use the closed forest for shelter and browsing during the dry season. The importance of game as a service, in seed dispersal and seedling establishment (in manure), must also be taken into account.

THE POTENTIAL CONTRIBUTION OF NON-TIMBER PRODUCTS TO TOTAL FOREST VALUE

To date, most tropical forest valuations have focused exclusively on timber resources. Although the economic contribution of timber is large, the value of all the biological resources obtainable from a given area in a tropical forest may sometimes be larger. The total value of a representative natural rainforest has yet to be estimated. Peters et al. (1989) took a 100 percent inventory of all the useful plants in a single hectare of forest 30 kilometers away from Iquitos, Peru, in which they measured density, derived a production function for each species, and calculated the market value of the products, as well as harvest and transportation costs from the site to market. The inventory showed that there were 60 timber species, 11 edible fruit species, and 2 latex-producing tree species in the forest plot examined. Forty-one percent of the species present in that one hectare of land yielded products with an actual market value. Sustainable yield of such resources might entail selective timber removal (30 cubic meters per harvest) on a 20-year cutting cycle with annual fruit and latex collection in perpetuity.

Using these criteria, Peters calculated that the native plant resources on his research site possess a net present value (NPV) of approximately $9,000. Surprisingly, 88.2 percent of the total NPV of the forest, as calculated by them, came from fruit and latex products. Though, as explained earlier, Peters's example is probably atypical, the NPVs he calculated demonstrate that natural forest utilization can be economically competitive with other forms of land use in the tropics. In comparison, the timber and pulpwood obtained from an intensively managed plantation of *Gmelina arborea*, admittedly on less fertile upland soil, in Brazilian Amazonia was estimated to have an NPV of $3,184 (Anderson 1983), or less than half what the example of naturally managed forest could earn. Gross revenues (without externalities such as fencing costs) from fully-stocked cattle pastures in Brazil on a variety of soils have been reported to be $148/hectares/year (Kahn 1988). The net present value of a perpetual series of such pastures discounted at 5 percent is only $2,960. Adding on the costs of weeding, fencing, and animal care would significantly lower this estimate.

Peters's study is one example of the type that should be carried out in different tropical forests in order to get a better estimate of the value added from non-timber goods that can accrue to timber-producing forests. One study cannot be considered representative of other forests, no matter how carefully executed, since forests will differ in species composition, soil fertility and moisture content, and canopy structure, hence light economy. Until such studies are completed, and more realistic estimates exist of the true value that non-timber products add to timber production, the non-timber-producing species should be incorporated into management protocols for the natural forest. It is certainly possible that such research will indicate that optimizing the value of undisturbed natural forest entails forgoing timber production altogether in favor of total genetic conservation or sustainable, nonintrusive production of non-timber goods.

INTERDEPENDENCIES

The record on non-timber goods reveals certain interdependencies:
1. Timber and non-timber goods can exist side by side in the forest. Timber trees can provide conditions (i.e., light shade and soil protection for plants; home and food for animals) favorable for development of other non-timber species. In some cases, though, as in Asian dipterocarp forests, they may be in competition for resources. Selective logging can positively influence some non-timber species. It has been shown that rattan grows best in the gaps in the forest created by logging, where some light is available for its development (see appendix 5.1). Mushrooms such as *Favolus brasiliensis* and *Polyporus tricholoma* typically grow from rotting logs and fallen trees. Since there is an abundance of fallen trees and rotting logs in logged-over forests, mushrooms might be cultivated in selectively logged forests as a part of forest management. Likewise, some non-timber producers are often early, successful species that also act as nurse crops for quality hardwood species. Many non-timber-producing species yield early in forest-regenerating areas following logging, and their productivity may increase in logged-over forests, assuming limited logging damage. In Sri Lanka, rattan, wild cardamom and other medicinal plants, and sugar palm experience earlier and higher yields following logging (C. V. S. and I.A.U.N. Gunatilleke, personal communication).
2. Many non-timber products are available from timber species with

multiple uses. Some examples from one forest in South America (Peters et al. 1989) include:

Hevea brasiliensis (rubber): latex, sawtimber
Spondias mombin (uvos): edible fruit, sawtimber
Couma macrocarpa: edible fruit, latex, sawtimber
Caryocar glabrum: edible nuts, oil, sawtimber
Parahancornia peruviana: edible fruit, latex, sawtimber
Brosimum rubescens: edible fruit, sawtimber
Manilkara guyanesis: edible fruit, sawtimber

3. Harvesting of timber and non-timber goods is not mutually exclusive. Timber harvesting occurs every 20 to 70 years, but non-timber goods are often collected annually and can provide a continuous source of income for local populations at times when timber is still growing. In the floodplain forests of Amazonia, fruiting of trees and adjacent fish populations, for example, are interdependent. Fish feed on the fruit and associated waste and insects; nutrients from fish effluents and mortality, in turn, are recycled back into the plant ecosystems of the floodplains (see appendix 5.2). The seeds of many trees are likewise dispersed by the fish, which consume various plant materials, transport the seeds, and through excretion and mortality, deposit them in new locations.

4. Timber harvesting may be deleterious to production of non-timber goods. Logging appears to have a negative effect on wildlife. Destruction of the habitat caused a decrease in bearded pig populations in Sarawak (see appendix 5.2). It has been estimated that, with current logging practices (see chapter 8), 40 percent of the residual stand may be destroyed (Marn and Jonkers 1981). Many of the trees destroyed would otherwise have yielded non-timber products. Under certain conditions, therefore, production of non-timber goods may justify ceasing timber operations altogether.

5. Some non-timber goods substantially affect the production of others. Illipe nut is a good example, because its irregularity of supply directly affects availability of labor for harvesting two other non-timber goods: bearded pigs and rattan (see appendix 5.1). Reproduction and survival of the bearded pigs is also influenced by illipe fruit, which are a preferred source of food. Leighton (1987, 6) noted that rattan prices and supply follow a seasonal trend. During the illipe nut fruiting season, which lasts a few months every three to four years, rattan collectors shift their attention to the more profitable nut. The decreased supply results in a 20 to 40 percent increase in rattan prices.

CONCLUSION

Non-timber forest products differ from timber in the large variety of species, the frequency and intensity of harvesting, the smaller yield per unit area, the higher value per unit weight, and, in some cases, the ease of synthesis. Their harvesting is labor intensive, but requires minimal investment.

Non-timber goods continue to be undervalued by policy makers. Although they are utilized locally, nationally, and internationally, and although they provide significant jobs, income, and materials for rural populations, their value is rarely reported in national and international statistics. Under certain conditions, the aggregate value of non-timber products can surpass the value of the tropical forest measured only in terms of timber or other wood products. Peters et al. (1989) demonstrated that in some forests in Latin America, the value of non-timber products can account for up to 88 percent of net present value of a hectare of tropical forest.

Non-timber goods from indigenous forests are neglected for the following reasons: (1) lack of adequate information concerning their economy and biology; (2) lack of established markets; (3) irregular supply and difficulty to maintain quality standards; (4) replacement by artificial substitutes and cultures in plantations; (5) lack of technology to process and store many of the perishable products; and (6) low net return to major trading interests.

Regional differences exist in the relative value of timber and non-timber products. While rattan and bamboo are extremely important to Asian forests, latexes, food crops, and fish are the most important products in the Latin American forests. African tropical forests are known for their wildlife. Many Asian forests are dominated by timber canopy trees that appear to compete with species yielding other goods. But cultivation of fruit in mixed species home gardens is a universal tradition in the tropics that reaches its greatest sophistication in Asia. In Asia in particular, the growing of hardwood timber plantations intermixed with producers of other products deserves consideration.

In the final analysis, although the available information is meager and insufficient, the added value of non-timber forest products can make mixed-use natural forest management a far more economically favorable option on agriculturally marginal land than timber production. Obtaining the necessary economic and biological information to test this hypothesis must be a first and major priority.

APPENDIX 5.1

Raw Materials for Craft Industries: The Case of Rattan

RATTAN REPRESENTS one of the most lucrative non-timber forest products to be considered in a mixed-use management plan for tropical forests. An economically vital commodity, rattan continues to generate significant foreign exchange for producing countries. Despite faulty attempts to increase the value added from rattan production and processing, Indonesia, the world's largest producer, could reap significant benefits from promoting cultivation of rattan within the context of mixed hardwood plantations.

THE ECONOMIC IMPORTANCE OF RATTAN

Until World War II and the onset of large-scale harvesting of virgin forest, rattan was the most important forest product in many Southeast Asian countries (Siebert and Belsky 1985, 522). At present, rattan is still the most important non-timber good in Southeast Asia. In Indonesia, rattan exports have increased in value a hundredfold from the early 1970s. In 1986, exports of raw rattan, matting, basketry, and furniture exceeded $100 million (see table A5.1). Thailand's exports of rattan furniture goods from 1983 to 1986 averaged about 430 million baht, or $15.5 million (Government of Thailand 1988).

In addition to generating export revenues, rattan is important at other levels as well. Rattan harvesting and processing is labor intensive and generates significant rural employment. According to one esti-

This appendix was prepared by Ricardo Godoy of the Harvard Institute for International Development.

Table A5.1

VALUE OF EXPORTS OF RATTAN FROM INDONESIA, 1981–1986

(IN MILLIONS OF U.S. DOLLARS)

Year	Raw and Semiprocessed		Matting and Basketry		Furniture	
	Value	%	Value	%	Value	%
1981	68.0	93	4.0	6	0.2	<1
1982	75.7	91	6.9	8	0.2	<1
1983	78.3	90	7.3	8	1.4	2
1984	86.0	91	7.2	8	1.5	1
1985	86.6	88	10.4	11	1.4	1
1986	84.1	81	14.0	13	5.7	6

BASED ON: Barichello, R. and R. Godoy, 1987. "An Assessment of Indonesia's Rattan Trade Policy," (unpublished monograph)

mate, the rattan industry in Asia employs half a million people in collection, cultivation, processing, and manufacturing (IDRC 1980, quoted in Siebert and Belsky 1985, 522). In 1985–86, for example, about 116,000 full-time workers were employed cutting rattan palms in Indonesia (Leighton 1987). Rattan harvesting and processing probably created employment opportunities for more people than this figure suggests, since these activities tend to be seasonal and part-time (Dransfield 1981, 180).

Unlike the price of many other non-timber forest products, rattan prices have increased dramatically in the recent past. Rattan has become a valuable resource; its real price in the world and Indonesian markets has risen at a real annual rate of more than 20 percent over the past 15 years. This trend of rising prices will probably continue in the immediate future owing to the shortages brought about by forest destruction and direct harvesting. Indonesia's recent ban on the exports of raw and semiprocessed rattans will add further impetus to the price rise.

Rattans are important to the local economy because they provide people with raw materials for binding, basketry, and matting. Furthermore, access to rattan provides rural people with a lucrative source of cash. Anecdotal evidence of several authors suggests that rattan cultivators tend to enjoy higher standards of living and income than noncultivators (e.g., Dransfield 1987a). Rattan harvesting also provides critical income to poor, rural households during lean agricultural years (Siebert and Belsky 1985).

AGRONOMIC ASPECTS OF RATTAN

Rattans are climbing palms found mainly in the dipterocarp forests of the Malayan archipelago (Moore 1973; Dransfield 1981, 179). Rattans grow well in virgin and secondary forests, especially in the gaps created by logging. The most productive species also require high moisture and soil fertility. Of the 600 species known, only about a dozen have commercial value (Manokaran 1984, 95); about a third of all rattan species are located in Indonesia. The economically most useful rattans include rotan manau (*Calamus manan*), rotan sega (*C. caesius*), and rotan irit (*C. trachycoleus*).

The thicker diameter canes (e.g., rotan manau) are used to build the frame of a piece of furniture. The thinner canes (irit or sega) are also used in the furniture industry for wrapping the low quality thick canes, for weaving the back, sides, and seats of furniture, and for making cores. The skin of thin canes is often used for making mats, mostly for local use or for export to Japan.

Rattans have woody, flexible stems that climb through the trees in the forests. Some have been known to reach 150 meters in length (Dransfield 1981, 183). Thickness varies from 0.3 to 3 cm (Purseglove 1972, 421–2). Rattans yield utilizable canes in six to seven years, but they do not come into full bearing until the fifteenth year. Mature rattans can have up to 50 or more stems, 26 to 30 meters long. Ten percent of these can be harvested every two to three years (Purseglove 1972, 421–2). Several multistemmed, slender species of rattan are also utilized. These types of rattan have the advantage of being suitable for cultivation, since stems of individual plants can be selectively harvested every two to three years once the plantation comes into bearing, without the cost that would otherwise be imposed by replanting and awaiting subsequent harvests. The wide-diameter canes are limited to a single harvest because of their single, unbranched stem. Several wide-diameter species with multiple shoots have been found in Sulawesi, Papua New Guinea, and Irian Jaya (Indonesia), but these are not the same quality as manau canes, nor have they ever been grown in plantations. Nonetheless, the Forest Research Institute of Bogor (Indonesia) recently identified two high-quality species in Sulawesi that have vigorous clusters, allowing more than one harvest. These species have been found to grow well in open spaces of disturbed forests (J. Dransfield, personal communication). There are also trial plots in Sri Lanka of two species with wide diameters.

LOCAL PROCESSING

Rattan is collected by teams of two or three people (Leighton 1987). Cooperative labor is required to pull down with force the long rattan stems, whose terminal whorls of spiny leaf tips are caught up in the canopy. Several people are also needed to carry the heavy bundles to collection points accessible by boats or other vehicles.

Once rattans of suitable lengths are discovered, they are cut and the stems are freed from the canopy by climbers. The search time for rattans in cultivated gardens is considerably lower than in the forest. In cultivated gardens, rattans are often mixed with rubber or fruit trees at a greater density than found in the forest. Consequently, the searching time can decline from 20 minutes/stem in the forest to less than 2 minutes/stem in a garden. Once on the ground, rattans are cut into suitable lengths.

To be durable, rattans must be scraped with sand, steel wool, or coconut fiber within a day after harvesting to remove the outer glassy tissue impregnated with silica. If scraping is not possible, rattans must be soaked in water, as deglazing of dry rattan is difficult (Cody 1983, 11). After scraping, canes are sun- or smoke-dried, depending on the season.

In processing centers, rattan is washed, fumigated with sulphur dioxide, or boiled in a mixture of diesel and coconut oil to expedite drying. Canes are then dried again in the sun to 5 to 10 percent moisture, sorted into different colors and thicknesses, cut, weighed, and bundled for shipment (Purseglove 1972). Canes processed up to this stage are considered raw.

After washing and fumigating, canes are often polished. The thicker canes are used to make furniture and, to a lesser extent, to make consumer articles such as walking, cricket, or polo sticks and drain-clearing equipment. The outer portion of the small-diameter canes is generally split; the peel or skin is used to weave mattings, baskets, bags, and other articles. The inner core is also split into flat or round wicks for weaving and binding. The core and peel are considered a semi-processed good.

FORMS OF CULTIVATION

Most rattans are harvested from the wild. Nonetheless, rattan is also cultivated as part of mixed gardens by sedentary cultivators, or is planted in burned-over forest by shifting cultivators in Indonesia. In Sarawak, Malaysia, rattan is cultivated near longhouses, mainly for household consumption (Dransfield 1987a).

In West Kalimantan and in Central Kalimantan, along the Barito River, Indonesia, some smallholders intercrop rattan with rubber (West Kalimantan) and other fruit trees (Central Kalimantan). Smallholders in Central Kalimantan have been cultivating small-diameter rattans (*C. caesius*; *C. trachycoleus*) for over a century in areas of low-lying secondary alluvial forest on the banks of the Barito River (Dransfield 1987a). These areas are subject to periodic severe and prolonged floods, rendering them unsuitable for cultivation of formerly more lucrative tree crops (e.g., coconuts, cocoa, oil palm), or for rice. Between Central and East Kalimantan, shifting cultivators plant thin-diameter rattans (*C. caesius*) after harvesting one or two years of dry rice in burned-over forest (Weinstock 1983, 1985; Weinstock and Vergara 1987). Rattans are ready for cutting after ten to fifteen years, just at the time cultivators are clearing the forest again to plant dry rice.

INTERNATIONAL TRADE IN RATTAN

Most of the world's supply of rattan is used in furniture manufacturing; the balance is used in webbing, basketry, and minor articles. Indonesia is regarded as the world's largest producer of rattan, exporting perhaps 50 to 70 percent of the world's cane requirements (Cody 1983, 19). Minor exporters include Malaysia, Papua New Guinea, the Philippines, and Thailand. To promote downstream processing, the Philippines and Thailand banned exports about 10 years ago and Malaysia imposed an export tax on raw materials. Thailand is a major producer, but it also imports from Indonesia, Myanmar, Laos, the People's Republic of China, and Malaysia. Most of Malaysia's rattan is exported to Taiwan, but smaller amounts reach the Philippines, Korea, and Japan.

Until the late 1970s, Indonesia exported primarily raw rattan. Half of its exports went to Singapore and Hong Kong, where rattan was processed and sorted before being re-exported to Taiwan, the People's Republic of China, Europe, and the United States. In the early 1980s, Taiwan and Europe began to purchase directly from Indonesia, circumventing Singapore. Nevertheless, 15 percent of Indonesia's rattan still finds its way to Singapore (*Business News* [Indonesia], June 3, 1987). Despite the cheaper prices gained by buying directly from Indonesia, foreign buyers often prefer to buy from traders in Singapore to obtain a wide assortment of canes for different ends and to ensure themselves of quality stocks.

The supply of rattan furniture is varied in quality and design. High-quality furniture is generally sold in Europe, Japan, and the Caribbean. This type of furniture is often exported unfinished, with final processing completed in the importing country. Furniture factories catering to the high-end market are found in France, Italy, Spain, Portugal, Germany, Holland, and the United Kingdom. During the 1980s, an increasing demand for medium- and low-quality furniture in the United States stepped up manufacturing operations in the People's Republic of China, Taiwan, and Thailand.

INDONESIA'S ROLE IN THE RATTAN MARKET

As the world's leading producer of rattan, Indonesia exercises a direct impact on the world rattan industry through its domestic and international trade policies. Since the late 1970s, the Government of Indonesia, like other Asian countries, has taken several steps to encourage downstream processing of rattan in order to increase employment, export earnings, and value added. In 1979, the Ministry of Trade imposed a ban on the export of unsorted, unwashed, unsmoked, unsulphured rattan. Nine types of raw rattan cane were subject to an export tax. In October 1986, the Ministry of Trade took an additional step in banning exports of sulphured, washed rattan. At the same time, it decreed that, until the end of 1988, the government would only allow the export of semiprocessed canes. These exports are now subject to a 30 percent tax. Finally, in July 1988, the Indonesian government banned the export of all raw or semiprocessed rattan products. These policies have generated several unforeseen consequences for the Indonesian government and for the harvesters of the resource.

1. *Higher Prices and New Suppliers.* The ban has depressed domestic Indonesian prices and raised world prices of rattan. The prospect of higher prices of raw materials has galvanized furniture manufacturers in Asia to search for new sources of raw materials. New commercial supplies of rattan have been located in Papua New Guinea, the Solomon Islands, and other countries. In response to the ban, many marginal suppliers are now assessing the feasibility of establishing rattan plantations.

2. *Possible Reduction in World Rattan Furniture Trade.* Rattan accounts for a small share of the world's total furniture trade. Wicker, willow, and rattan together account for only 0.4 percent of the total U.S. furniture

market (Mei and Sieh 1985, 32). Although demand for high-quality rattan furniture is relatively price inelastic—price increases do not reduce demand—medium- and low-quality rattan furniture face much competition from substitute of other materials. Since the lower grades of manufactures represent 80 percent of the value of the international trade in rattan furniture, a price hike in raw materials is likely to encourage both producers and consumers to switch to substitutes.

3. *Reduced Supplies.* The world supply of rattan in the natural forest is being rapidly depleted by logging, forest destruction, and direct harvesting. Current natural supplies in Indonesia may only last another 15 years. The rate of collection increased dramatically as a result of Indonesia's ban: in the few months before the official decree of a ban, foreign buyers began hording raw materials in anticipation of the moratorium on exports, thereby accelerating the rate of harvesting. At present, about a third of rattan species in Malaysia and Indonesia are under the threat of extinction (Dransfield 1987a).

PROSPECTS FOR CULTIVATION WITH MIXED HARDWOODS

Preliminary evidence suggests that rattan cultivation on a large scale may be financially viable. One recent study from Malaysia suggests that rattan cultivation yields a rate of return of about 6 percent (Noor and Razali 1987), a deceptively low figure for three reasons. First, the plot selected for the study was not optimal for rattan cultivation, since it was steep, well drained, and displayed a dense canopy. The returns would have been much higher on a better site. Second, a 6 percent return for rattan compares favorably with the financial returns of monocropped traditional tall variety coconuts in favorable growing conditions (e.g., North Sulawesi, Indonesia). The cultivation of such coconuts in North Sulawesi yields financial returns ranging from 4 to 7 percent, depending on the management level and the specific agro-ecological conditions (Godoy and Bennett 1987). Third, rattan requires arboreal support, so that rattan must be grown in mixed culture with timber or other trees, which yield an additional return.

Given the potential returns on investment and benefits from intercropping with other species, production of rattan in combination with stands of mixed hardwoods could provide Indonesia with the means to promote sustainable production of both timber and non-timber forest products in suitable regions. Indonesia would be well advised to launch a full cost-benefit analysis including variable returns from rattan production juxtaposed with various timber-harvesting techniques.

APPENDIX 5.2

The Added Value Contribution from Wildlife

AMONG THE most valuable non-timber products in tropical forests is wildlife, which may be harvested for direct export, local consumption, or industrial processing (see *Raw Materials for Industry*). Two poignant examples of the significance of wildlife, ecologically and economically, are the bearded pigs of Sarawak and Amazonian fish.

BEARDED PIGS IN SARAWAK AND MALAYSIA

Bearded pig (*Sus barbatus*) is by far the most hunted animal in Sarawak. It contributes well over 80 percent of all game (Caldecott 1987). Other animals frequently hunted in the region are pelandok (mouse deer, *Tragulus*), kijang (barking deer, *Muntiacus*), and rusa (sambar deer, *Cervus unicolor*). It is estimated that nearly 20,000 metric tons of wild meat is harvested every year in the state as a whole. This is equivalent to an average consumption of about 12 kg/person/year, though this varies greatly among different areas. All the meat that is not locally consumed is sold in nearby towns. The trade can be highly profitable because of low transport costs and high price differentials between source and market. For example, bearded pig meat can be bought for $0.50 to $1.00/head upriver and sold for $6.00/kg in the coastal city of Sibu, having incurred a transportation charge of only $0.33/kg (Caldecott 1987, 49).

Besides its value as a source of food or a source of cash, if traded, bearded pig is a source of self-employment for many rural people—as hunters, middlemen, or traders. It has been estimated that, on average, 18.5 man-days/year/person are spent in hunting by Sarawak villagers (Caldecott 1987, 59).

It is easy to think of wildlife in economic terms when considering the products being bought and sold for cash. It is more difficult to appreciate the economic value when the wild game is neither privately owned nor traded in the cash economy. Measured in terms of its replacement cost, through a state-wide livestock development scheme, the inferred current value of wild game in Sarawak would total $55 million/year (Caldecott 1987); if fishponds were created in the state to substitute for game products, the replacement cost would total $41 million/year.

In economic terms, the bearded pig harvest presently supplies high-quality food at very low monetary cost to the cash-poorest groups in the state. With the population increase, a strategy is urgently needed to manage the bearded pigs on a sustainable basis or even increase their numbers. Such a strategy should incorporate several factors.

First, growth, maturation, and interbirth interval appear to be directly related to fruit availability—especially illipe nuts, acorns, and chestnuts, and all oil- and fat-rich seeds (Caldecott 1987). As the fruiting intensity of these plants varies greatly from year to year, so does the number of the bearded pigs. Therefore, protection of the food supply is at least as important as general habitat protection.

Second, the viability of the bearded pig population is directly related to the timber industry. Caldecott demonstrates that reduced populations, measured in declining yields of meat per unit of hunting effort, are associated with logged regions of tropical forest. Hunting in logged areas produced an average of 1.1 kg/man-hour compared to twice that amount—2.2 kg/man-hour—in unlogged forests. Caldecott showed that there is typically a sharp decline in wildlife in the years subsequent to logging (see table B5.1). Whether the effect of logging on bearded pig populations is due primarily to habitat destruction or to improved access for hunting by outsiders is unclear: since the 1940s the rural population of Sarawak has more than doubled, and the number of shotguns owned by them has increased by at least 50,000.

Finally, and most importantly, the relative costs of replacing the food supply through other means must be measured against the benefits of a sustainable logging industry. Even when not traded in local or international markets, non-timber forest products can constitute highly valuable assets.

Table B5.1
MEDIAN NUMBER OF ANIMALS KILLED IN LOGGED AREAS
(PER 10 FAMILIES PER YEAR)

Type of Animal	Unlogged	Number of Years Since First Logging		
		1–10	11–20	21–30
Bearded pig	99.7	32.1	13.5	3.4
Kijang	7.1	1.0	0.4	0.3
Pelandok	6.8	9.1	2.3	1.9
Rusa	2.5	1.2	0.8	0.5
Weight in meat (kg)	3,806.0	1,240.0	534.0	155.0
Daily ration (g)	149.0	49.0	21.0	6.0

BASED ON: Caldecott 1987.

AMAZONIAN FISH

The undervalued non-timber goods in the Amazonian basin consist not only of indigenous food crops and latexes but also fish. A large portion of the commercial catch of fish from Amazonian waters is represented by taxa that, as adults, are sustained on fruits, seeds, insects, and detritus derived from flooded forests of the clearwater and low-productivity rivers.

A study on total yield from fishing within a 60-km radius of Itacoatiara, Brazil, estimated the productivity of the Amazon flood-plain and the per capita availability of fish (Smith 1981, 83). The total catch from the study area for 1977 was estimated to be 3,151 tons. Since the average width of the Amazon floodplain within a 60-km radius of Itacoatiara is about 30 km, the harvested yield of fish was 0.9 tons/square kilometer (Smith 1981, 86). About 50,000 people live in the study area; thus there was sufficient fish captured to supply each person with 104 grams/day, allowing a 40 percent loss due to discarded portions and not counting catfish, which are not eaten locally but are exported. Since an average person needs only 35 to 40 grams of animal protein a day (0.6 g protein/kg body weight), 104 grams is more than enough fish to supply the protein needs of the inhabitants in the region, not accounting for seasonal variation in the catch.

When compared to alternate sources of protein—wild game or cattle—fish protein is significantly less expensive. Although beef is readily available in Itacoatiara, the cheapest cuts from cattle are at least three times more expensive per unit weight than medium-priced fish.

Fishing is more productive than hunting in the nearby undisturbed forests: the game yield from the 200-square-kilometer area of upland forest yielded 33 kg/square kilometer, whereas fishing in the Amazon *varzea* was about 27 times more productive (900 kg/square kilometer). These numbers, from an ecological and management standpoint, are not directly comparable considering the fact that fish benefit from nutrients derived from the whole catchment area, while the upland forest game relies only on resources within the immediate region. Both meat sources, however, depend on the sustained management of the forests of the whole catchment for their own survival.

The rural population of the Amazon valley, due to local beliefs, never eat certain kinds of fish, particularly catfish. In this category are Pariba, Amazonia's largest fish, weighing as much as 136 kg (Myers 1947); Pirarara, which weighs up to 80 kg; and Aruana, whose light meat, with low fat content, is considered exceptionally easy to digest (Smith 1981, 91). Most of the catch of these species is exported.

For purposes both of export and local consumption, therefore, the non-timber resources of the tropical forest in the Amazonian floodplain must be included in the opportunity cost of logging and timber production.

CHAPTER 6

Environmental Services: Another Major Component of Forest Value

TROPICAL FORESTS provide important benefits in the form of environmental services that are often necessary for sustainable economic development. These services are functions performed naturally by forest ecosystems, such as regulation of droughts and floods, control of soil erosion and sedimentation of downstream water bodies, amelioration of climate, barriers against weather damage, and groundwater recharge. Benefits derived from the rich biological diversity and gene pool found in tropical forests also count among the important environmental advantages.

Despite the significant economic contribution of forest services, they are often undervalued, primarily because they are not priced in the marketplace, and because the full effects of their destruction often are not realized until after the short-term benefits from their destructive use have been enjoyed. Forests can be viewed as natural capital, providing a perpetual stream of benefits and services that support, enhance, and protect economic development and the quality of life. The replacement costs of lost services due to the ill-advised or inappropriate destruction of important forest capital can be very high. The failure to adequately value and protect the services provided by the natural forests today greatly increases the capital and other costs of economic development in the future.

In many cases the revenues generated or recovered by logging forests may be lower than the value of the stream of services provided by leaving the forests in their natural state. Serious efforts to identify key watershed areas, study their dynamics and values, and incorporate the information and implications into development planning and management are still quite limited in most countries. Future benefits from forests maintained in their natural state as gene pools and sources of germplasm are not easy to calculate. Nevertheless, they should not be ignored because new discoveries and innovative uses of forest products with high economic and other value occur regularly. In key watersheds in particular, logging, if allowed, should be strictly regulated and managed so as to protect the forest's environmental functions.

Logging may have unforeseen but potentially disastrous environmental consequences that can detract from the future viability of timber production. Ill effects may arise from logging roads that open new areas to encroaching settlers who then convert the forest to other uses. Large quantities of potentially combustible debris are usually left on the ground as a direct by-product of logging activities. Pioneer tree species that establish themselves on logging roads also produce increased amounts of combustible litter. Excessive debris seems to have been a contributing factor to the widespread destruction of timber and other resources caused by the extensive fires in Borneo in 1983 (Leighton and Wirawan 1986). Such fires, if repeated, eventually cause wastelands and, of course, encourage the growth of combustible grasses such as alang-alang. Soils become compact and prone to gully erosion. Restoration to any form of productive use requires significant capital investment.

Environmentally and socially disruptive logging methods may result in strong local or international condemnation, and ultimately to constraints on future logging, whether justified or not. The same consequences can result from the establishment of plantations or reforestation projects. In Karnataka, India, controversy surrounding the widespread planting of eucalyptus trees by the government resulted in one leader of the powerful Farmer's Association threatening that "if the government does not accede to our demand [to stop planting eucalyptus], we need only two days to ensure that eucalyptus is wiped out of Karnataka" (Center for Science and Environment 1982). Similar conflicts arose in Thailand recently when the government began issuing leases to large-scale investors to establish eucalyptus plantations in encroached forests.

A number of claims or assumptions held about the relationships between tropical forests and water flows, drought, floods, and precipitation form the basis for forest watershed policies in developing countries (Hamilton 1983, 123–31). Currently, most research from watershed experiments comes from temperate, industrialized countries, which may not be valid for tropical regions:

> Direct extrapolation of results to different combinations of climate, soil, and vegetation, even within the temperate zone, always has been problematical, so we should not be surprised that direct transfer of research to an entirely different climatic zone is fraught with uncertainty. There is no reason to believe that the basic processes are different. The research to prove it, however, must be carried out under a wide range of conditions so that a set of generalizations can be made for the range of conditions that exist in the tropics. (Hamilton 1983)

Large gaps remain in current knowledge of the effects of conversion or degradation of tropical forests on watershed processes. Circumstantial evidence appears to suggest clear cause-and-effect relationships, for instance between logging in headwaters and floods downstream. However, one review of evidence concluded otherwise (Hewlett 1982, cited in Hamilton 1983, 127):

> [Degradation, on the other hand,] encompassing whole river basins, may indeed aggravate flooding and be one of the principal causes of serious flood damage. However, if converted to controlled grazing lands or agriculture under a sound soil and water conservation regime, such watershed land use should no more cause floods than would careful forest harvesting.

These caveats should be borne in mind when considering some of the more general statements made in the following sections. Hamilton (1983, 126) poses basic questions about the degree to which "proven" relationships between logging and floods are real or a convenient but unfounded explanation for natural flooding, exacerbated by the effects of population. It is the extent to which the benefits of environmental services are mitigated by the benefits from other forest uses, including timber production, that must constitute one factor in a cost-benefit analysis of multiple-use management.

ECONOMIC BENEFITS OF ENVIRONMENTAL SERVICES

Sectors dependent upon the environmental services provided by forests include agriculture and fisheries, energy, public health and water supplies, transportation, and tourism. Tropical forests yield significant benefits for each of these sectors, as well as for national and international society as a whole in the form of climate moderation, carbon sequestration, mitigation of natural disasters, and harboring of unique genetic resources. Although not priced by markets, these benefits must be factored into a calculus of timber extraction to determine the true opportunity costs.

Agriculture and Food Production

Approximately 40 percent of farmers in developing countries live in areas that are dependent upon watershed functions provided by forests (World Bank 1987). Likewise, agricultural exports worth $36 billion a year are dependent upon forest-generated or protected soils and water (Clay 1982). It has long been recognized that forests retain large quantities of water, slowly releasing it over time, thereby mitigating flood damage in seasons of heavy run-off and preserving supplies for drier periods. Downstream farmers who practice year-round agriculture are particularly dependent upon uninterrupted supplies of irrigation water. In a study of the Tai forest in the Ivory Coast, it was reported that rivers flowing from primary forest release anywhere from two to five times as much water as do rivers from a coffee plantation zone (Dosso et al. 1981, cited in Myers 1984).

When large areas are deforested, heavy rain and floodwater can cause erosion of slopes and of productive soils (Marsh 1864; Jacks and Whyte 1939; Morgan 1979). Wiersum (1984) gathered much of the existing research on erosion of various types of forested and cultivated land (table 6.1); comparison of erosion rates indicates that ground litter and ground cover, as well as the presence of natural or plantation forest, constitute important factors in surface run-off and erosion. Soil erosion throughout the Pacific drainage areas of Central America, where large areas have been denuded of trees, is so serious that up to 40 percent of all lands along the Pacific slope are experiencing a decline in their productive potential (Leonard 1986, 169).

Forests reduce rates of sedimentation in irrigation canals and reservoirs. Much of the expected increase in agricultural production worldwide will have to come from irrigated croplands. Due to sedimentation from deforestation, the capacity of India's Nizramsagar Reservoir has been reduced from almost 900 million cubic meters to less than 340 million cubic meters, resulting in a water deficit for irrigation of the 1,100 square kilometers of rice and sugar cane for which the reservoir was created (IUCN 1980b, cited in World Bank 1987). On the island of Java, which presently retains only 15 percent of its original forest cover,

Table 6.1
EROSION IN VARIOUS TROPICAL MOIST FOREST AND TREE CROP SYSTEMS
(TONS/HECTARE/YEAR)

	Minimal	Maximal
Multistoried tree gardens 4 locations, 4 observations	0.01	0.14
Shifting cultivation, fallow period 6 locations, 14 observations	0.05	7.40
Natural forests 18 locations, 27 observations	0.03	6.16
Forest plantation, undisturbed 14 locations, 20 observations	0.02	6.20
Tree crops with cover crop/mulch 9 locations, 17 observations	0.10	5.60
Shifting cultivation, cropping period 7 locations, 22 observations	0.40	70.05
Taungya cultivation 2 locations, 6 observations	0.63	17.37
Tree crops, clean-weeded 10 locations, 17 observations	1.20	192.90
Forest plantations, burned/litter removed 7 locations, 7 observations	5.92	104.80

BASED ON: Wiersum 1984. Reprinted by permission of the East-West Center.

croplands lose topsoil each year equal to that required to grow rice for approximately 11.5 to 15 million people (Daryadi 1981; Soemarwoto 1979; Sumitro 1979, cited in Myers 1984, 384). Preventing the conversion of certain key forests to other uses can play a crucial role in supporting agriculture (Rockefeller Foundation 1981, cited in World Bank 1987).

About 500 million people live in the Ganges River Plain of India and Nepal, the principle grain-growing area of both countries. The forest cover of the upper catchment areas has been reduced by at least 40 percent in the last thirty years, contributing to greatly increased damage from monsoon flooding followed by drought. While flood damage used to cost approximately $120 million each year before 1970, costs now reach $1 billion to $2 billion (Center for Science and Environment 1982). Flood control, in the form of embankments, dikes, and large dams, now costs India about $100 to $250 million each year, while only negligible amounts are spent on forest conservation (Myers 1984, 263; Spears 1982). Considering the loss of crop production and soil nutrients due to flooding, and increased soil erosion due to deforestation, one researcher estimated that the annual value of India's watersheds is $72 billion and that it would cost nearly $48 billion to construct even earthwork reservoirs to store the same volume of water that the forests soak up and release each year (Ranganathan 1978, cited in Grainger 1980).

Virtually every major watershed in Central America is suffering serious devegetation and erosion, which disrupt the water cycle and deposit extremely high loads of soil sediments into streams and rivers. Annual flooding in several of the major valleys in Honduras has increased dramatically in recent years, causing an average loss of $50 million in damage to crops and infrastructure. The disastrous losses from Hurricane Fifi in 1974 are believed to have been greatly exacerbated by the reduced carrying capacity of the heavily silted streambeds and the greatly increased peak waterflows (Leonard 1986, 181–82). In Amazonia also, flooding and damage due to deforestation appear to be worsening (Myers 1984, 264).

Tropical forests also benefit agriculture by serving as "baseline" scientific study areas that can provide information on nutrient cycling, energy flow, species interaction, and other natural ecological processes that can serve as models for sustainable agricultural systems. In addition, the tropical forests serve as refuges for plant and animal species, thereby sustaining agricultural and livestock production, as is the case with several insects that act as crop pollinators (World Bank 1987).

Fisheries, too, rely on forests both directly for nutrients and breeding grounds (see appendix 5.2) and indirectly for soil and water regulation. Following deforestation in the Philippines, silt deposits far offshore suffocate coral reefs and mangrove ecosystems, thus impoverishing valuable fisheries. It is estimated that 75 percent of all the fish sold in Manaus, Brazil, which provides the largest market in western Amazonia, depend ultimately upon the seasonally flooded *varzea* forest. This is because the fish are sustained by the fallen fruit, insects, and other organic detritus dropped into the water from the vegetation (Goulding 1980). Between 1970 and 1975, the fish catch in Amazonian rivers fell 25 percent after deforestation of the fishes' breeding grounds had occurred (Soulé 1982, cited in Caufield 1985).

Energy

The importance of forests in protecting hydroelectric investments is well recognized and there are numerous examples of heavy economic losses due to high loads of sedimentation that reduce generating capacity and significantly shorten the useful economic life of dams. Costa Rica, for instance, depends upon hydroelectric facilities for 99 percent of its electricity, but watershed deterioration as a result of deforestation is threatening the future of almost every major dam. Sedimentation at the Cashi dam is estimated to have resulted in lost revenue equal to between $133 million and $274 million for the project, which is barely 20 years old (Leonard 1986, 180). This may affect Costa Rica's plans to export excess electricity to its neighbors.

The 80,000-kilowatt Tavera dam project, the Dominican Republic's largest, was completed in 1973. By 1981, 18 meters of sediment had accumulated behind the dam, reducing its dead storage capacity by 40 percent, and active storage by 10 to 14 percent (JRB Associates 1981).

The Tehri Dam in India, one of the highest in the world, as well as an extremely expensive project, was planned to last 100 years. However, due to massive deforestation and erosion in the Himalayan foothills, the Indian Geological Survey recently reduced the dam's estimated life to between 30 and 40 years (McDonald and Reisner 1986, 294).

The Ambuklao Dam in the Philippines is silting up so fast that its useful life is being reduced from its planned 56 years to 32 years because of deforestation of the Agno River watershed (Myers 1984, 272).

In Ecuador, the Poza Honda Reservoir, built in 1971 and planned to last 50 years, is losing its capacity at a rate that will render it useless within 25 years. However, it was estimated that a conservation program to reforest that half of the 175-square-kilometer watershed which has lost forest cover would cost only $1.8 million and extend the reservoir's life to its planned life-span, producing at least $30 million worth of benefits (Fleming 1979, cited in Myers 1984).

Public Health and Water Supplies

While it is clear that forests regulate waterflow, the effects of deforestation on waterflow and the water table are not uniform, given variations in local topography, soil content, species composition, and the proximity of the water table to the surface (Hamilton 1983; Myers 1984). By assuring dependable supplies of potable water, tropical forests benefit public health and certain types of industrial output, in both rural and urban areas. In some rural areas, deforestation causes groundwater sources or springs to become dry, forcing people to drink water from polluted streams; in Bangkok, disrupted waterflow attributed to deforestation upstream has forced the city to turn to groundwater supplies and has contributed to the city's gradual sinking (Myers 1984,

275–76). At the same time, upstream deforestation in Haiti contributes to the clogging of sewers in Port-au-Prince (Ledec 1985).

It is ironic that some of the wettest areas in the world suffer from water shortages during various times of the year due to deforestation. Water is periodically rationed in Kuala Lumpur, Manila, Lagos, Abidjan, and Bangkok, and in other urban areas in the tropics water shortages frequently occur as a result of disrupted river flows (Myers 1984).

Transportation

Watershed functions provided by forests also support the transportation sector in developing countries. When forest cover is removed, the resulting heavy erosion and/or floods can bury roads. Sedimentation of harbors, canals, streams, and rivers can make them unnavigable. Siltation in the Ganges River system has rendered some sections of the main river impassable, while for several months of the year some downstream industrial installations suspend activities due to lack of water. The ports of Calcutta and Dacca are also silting up (Myers 1984, 271). In Argentina in the early 1970s, authorities were spending $10 million a year to dredge silt from the Plata River mouth and keep the port of Buenos Aires open to ships. A small portion of destroyed watershed can also have major consequences: 80 percent of the sediment in the Plata River came from only 4 percent of the drainage basin—the overgrazed Bermejo River 1,800 miles upstream (Pereira 1973, cited in World Bank 1987).

The deforestation of the Panama Canal watershed is leading to the silting up of the canal; heavy sedimentation and inadequate amounts of water during some dry seasons have already led to some cargo ships being diverted around Cape Horn, or to cargo to being unloaded and carried by rail to the other side of the canal (Timberlake 1987, 30).

Tourism

The tourist industry in many tropical countries consistently ranks among the top foreign-currency earners, often outranking wood-product exports. Nature-oriented tourism in particular is growing at a fast pace, and the tourists are often scientists and researchers conducting long-term studies. The Organization for Tropical Studies, a consortium of U.S. universities and institutions researching the tropical forests of Costa Rica, generates between $3 million and $10 million annually or 2 to 3 percent of Costa Rica's national tourist receipts (Laarman 1987). Nature-oriented tourism is fairly well developed in Costa Rica, but the potential is large for many tropical countries.

Thousands of foreign visitors annually contribute more than $200,000 to Rwanda's economy through park entrance fees alone, many trying to view some of the few remaining mountain gorillas in the world. At the same time that the forests sheltering these gorillas are preserved and earning foreign currency, the forests are also helping to maintain a balanced hydrologic regime and thereby sustain downstream agriculture (Weber and Vedder 1984). The mangrove swamps of Morrocoy National Park in Venezuela attract 250,000 to 500,000 visitors annually (Hamilton and Snedaker 1984, cited in OTA 1987).

Bird-watching, bird-feeding, wildlife photography, and general wildlife observation in the United States generated expenditures of $7 billion to $15 billion in 1980 (U.S. Fish and Wildlife Service 1981, cited in World Bank 1987). According to the World Wildlife Fund (1982, cited in World Bank 1987) 245 out of 645 breeding North American bird species are migratory species that live part of the year in Latin America. Thus, tourism in some temperate countries also benefits from the maintenance of tropical forest wildlands.

As interest in nature and outdoor recreation continues to grow in both developed and developing nations, tropical nature tourism and pristine forests will become increasingly valuable. Nature tourism has specific characteristics that make it particularly appealing to some countries: (1) nature-oriented tourists stay in the country longer than other types of travelers, and (2) nature-oriented tourism distributes the economic benefits of tourism to both urban and rural residents whereas traditional tourism tends to focus on urban areas (Durst 1987).

Moderation of Climate and Atmosphere

Tropical forests have important effects on climate at both local and regional levels. The exact dynamics are not well understood, and interpretations of available data conflict. Nevertheless, some effects are evident.

Because the forest of the Amazon basin generates about 50 percent of its own rainfall, there is evidence that extensive deforestation could trigger an irreversible drying trend in the region (Salati 1981, cited in World Bank 1987). Extensions of the dry season and increased flooding have already been recorded in the Manaus area (Stone 1985, 153).

Deforestation seems to be having an effect on the Panama Canal watershed. Research shows that rainfall at both ends of the canal has been fairly constant over decades, but meteorological stations inland have been showing a decline in rainfall of one inch per year over a number of years (Timberlake 1987, 30). As mentioned above, this process has already created limitations on the use of the canal in some dry seasons.

Many scientists express concern that clearing extensive patches of tropical forests will increase the reflectiveness (albedo) of the earth's surface, which could effect major alterations in global patterns of air circulation, and thereby shift rainfall distribution (Dickinson 1981; Potter et al. 1975; Potter 1981; and Sagan et al. 1979, all cited in Myers 1984). Clearing and burning of tropical forests release significant amounts of carbon into the atmosphere, which may be contributing to a global heating of the planet via the buildup of CO_2, a major greenhouse gas (Houghton et al. 1983 and Woodwell et al. 1983, cited in Myers 1984). Agricultural soils and, in particular, irrigated land created by forest conversion release other greenhouse gases, including methane and carbon monoxide, and nitrogen oxides that catalyze ozone production, several orders of magnitude more than forest sods. Nitrogen oxides also react with atmospheric water to form nitric acid, a major component of acid rain. Forests in leaf, including tropical evergreen forests throughout the year, are net absorbers of these gases, including ozone and the biogenic greenhouse gas neoprene (Keller et al. 1990). The buildup of CO_2 could theoretically have positive impacts, such as stimulating photosynthesis, thereby increasing agricultural productivity and creating additional plant biomass that would utilize much of the carbon added to the atmosphere each year, but that this would happen is by no means clear. Negative effects could include the melting of the polar ice caps and consequent rise in global sea levels and flooding of coastal cities (Woodwell et al. 1983; Niehaus 1979). Much uncertainty remains concerning the data, however, and controversy continues over what the consequences may be.

Carbon Sequestration

Although forests managed for timber production and timber plantations are a major sink for atmospheric carbon, natural unharvested forests are in near equilibrium and release as much carbon dioxide through decay as they fix through photosynthesis. A specialized form of global climate regulation occurs in the form of carbon storage through photosynthesis; just as tropical forests release large amounts of CO_2 when they are burned, they store large amounts in their natural state.

Estimates of the value of this particular environmental service vary considerably. The value of sequestered carbon on an undisturbed hectare of Amazonian forest, for example, is estimated to range from $374 to $1,625, while land prices range from $20 to $300/hectare (World Bank 1991). Assuming the marginal cost of emission reduction in the developed countries averages $28 per ton (Krutilla 1991; Nordhaus 1990), the carbon storage function of the world's closed tropical forests, in service replacement cost terms, is not insignificant. Among twenty countries, each with forests ranging from 3.5 million hectares to 375 million, the values range from $500 million to $1.3 trillion; the total current value over the 20 countries is more than $3.7 trillion (Panayotou 1992, 5).

Although scientists and a very few policy makers are only beginning to consider the economic implications of this particular environmental service provided by tropical forests, the potential benefits of such recognition are enormous once policy makers are able to internalize them (Panayotou 1992). Krutilla (1991) has estimated that the present value of carbon sequestered by the Malaysian forests ranges from $2,950 to $3,682/hectare, and may well exceed the net present value of timber, disregarding all other non-timber goods and services provided by the forests.

Cyclones, Earthquakes, Landslips, Etc.

Many so-called natural disasters may actually be caused or exacerbated by man (Wijkman and Timberlake 1984). Events that would otherwise be minor can cause considerable damage and loss of life when ill-advised clearance of strategic forests occurs. Deforestation can aggravate landslips and rockfalls, and increase damage caused by earthquakes and cyclones.

The adverse effects of torrential rains, strong winds, and storm surges associated with cyclones are reduced by the buffering effect of forests, particularly in coastal areas. A case in point are the catastrophic landslides and floods of November 1988 in southern Thailand, the worst natural disaster in Thailand's recorded history: over 350 people were killed and 30,000 displaced. Estimates of damage to crops, property, and infrastructure range from $200 million to $300 million. The disaster was attributed to a combination of high rainfall and the low natural resilience of steep-sloped granitic mountains, which has been further reduced by human economic activities. Among those activities the most important factor contributing to the event was the replacement of natural forests on the slopes with young rubber plantations without land conservation measures. The high rainfall and the reduced resilience of the area "collided" to produce the November landslides, debris flows, and the downstream floods (see table 6.1 for examples of erosion rates on plantations with reduced ground cover versus natural forests).

Biodiversity Values

Tropical forests should be perceived as major genetic resources because of the extremely rich biological diversity that characterizes them (see chapter 10). Biological diversity refers to three elements: the variety and number of different ecosystems found in a country or region, the number of different species and their relative frequencies, and the genetic variation within each species. Although preserving biological diversity can be justified scientifically, aesthetically, ethically, and economically, the focus in this chapter is on economics.

Diversity is necessary for several economic reasons: (1) to sustain and improve agriculture, (2) to provide opportunities for medical discoveries and industrial innovations, and (3) to preserve choices for addressing as yet unforeseen problems and opportunities for future generations (OTA 1987). Genes transferred to domestic crop plants from their wild relatives can increase yields, improve quality, provide resistance to pests and diseases, extend growing ranges, and permit wide hybridization between crop species or between them and related wild species. Maintaining genetic diversity is important to sustaining and increasing agricultural productivity. Most crops have been developed from a limited number of genotypes. Periodically, it becomes necessary to locate new germplasm in order to improve crop characteristics, and increase productivity and resistance to pests and diseases (see chapter 10).

The previous chapter described the enormous variety of useful industrial raw materials, foods, and other goods that tropical forests provide. The potential for new uses, products, and markets for tropical forest materials is evident given that (1) most species in the tropical forests have not been scientifically studied—let alone discovered and named (Myers 1984); (2) some of the world's most important and commonly used products originated from tropical forests; and (3) some of these products only developed regional or international markets relatively recently (Vietmeyer 1975). The National Academy of Sciences (1982) states that "although it is doubtful that any new major crop plant will be found in Europe or North America, it is entirely likely that one or more might be found in the humid tropical forests." It was discovered that *Copaifera langsdorfii*, a tree that only grows in northern Brazil, produces a sap that can be used directly as fuel in diesel engines (IUCN 1980a, cited in World Bank 1987).

The history of the rubber industry illustrates a number of points. The industrialized world did not even use rubber until the nineteenth century. In 1986, exports of natural rubber earned $3 billion for tropical countries (IMF 1987). The natural rubber industry literally grew from a mere 22 seeds collected from a single area (Carpenter 1983, 20). Tree breeding and selection technologies have resulted in enormous productivity increases in recent years and will continue to do so (World Bank 1987). However, it was recognized that reliance on such a narrow genetic range led to "genetic erosion" and loss of available germplasm. Consequently, the industry has been experimenting with other wild rubber provenances that may again increase productivity and also provide resistance to the South American leaf blight (Prescott-Allen and Prescott-Allen 1986). Similarly, cacao, coffee, sugar cane, and pineapples, among other crops, stand to benefit from current research and use of new germplasm collected from the wild. Crop genetic improvement has been responsible for about 50 percent of the total productivity increase, worth $1 billion in annual earnings, in the U.S. agricultural sector (OTA 1987). The foreign exchange value of oil palm in Malaysia increased by $57 million in the first year after its pollinator, an African weevil, was introduced from the forests of Cameroon (World Bank 1987).

Logging, even when carried out on a highly selective basis, may have negative consequences on the diversity of biological resources. However, these negative effects can be minimized by managing logging activities to take account of the biological characteristics of tropical forests, such as:

1. their species-area relationships (larger regions tend to have more species than small sites);
2. their competitive interactions (introduction of exotic species or varieties into areas where the species has no natural competitors, particularly following a disturbance such as logging, can lead to infestations);
3. the narrow endemism (restricted geographic range) of many rainforest species;
4. the high species richness (number of species) of most tropical evergreen forests;
5. the species interdependencies that are often intricate in tropical forests; and
6. the natural vulnerability of many species to forest disruption.

Chapter 10 will address the relevance of these concepts to timber production.

Environmental Effects of Forest Modification and Conversion

Conversion and alteration of forests may or may not be incompatible with the provision of environmental services, depending upon the type of change, the intensity and extent of the activity, and the topography and morphology of the region in which the changes occur. Furthermore, even when there is some trade-off between an economic activity and environmental services, some decline in the environmental services may be considered acceptable given the other benefits derived from the new use of the land.

Hydrological studies of catchments have rarely been conducted in regions such as the tropical high forests where mean monthly precipitation exceeds evapotranspiration for most of the year, where soils are generally water-saturated, and where windlessness and high relative humidity reduce evapotranspiration from forest canopies. In such areas, the expected increase in water entrapment and soil water storage, achieved through reduction of evapotranspiration, may rarely be realized and may be more than offset by: (1) soil surface compaction and dehydration leading to lower water infiltration; (2) the costs incurred by increased soil erosion and siltation; and (3) the effects of increased albedo, which presumably affects regional weather patterns (further research is required to determine the nature of these effects).

Moreover, it is common to hear general statements on the role of forest alteration in causing major floods, that reforestation or afforestation will increase streamflows. These statements are all too often solely based upon assumptions, circumstantial evidence or correlation, political expediency, or ignorance. Some of this common wisdom actually contradicts what is known about forests and watersheds. Conversion, whether for shifting agriculture, annual cropping, food or commodity tree cropping, or agroforestry, or alteration through commercial logging, fuelwood harvest, or plantations, leads to various effects depending on the nature of the forest and the type of intervention.

Although foresters and land-use planners often blame shifting agriculturalists as the principal villains in watershed alterations, Hamilton does not share this opinion (1983, 20):

Much of the above discussion is taken from the synthesis of discussions and papers from a workshop on effects of forest alterations convened in 1982 by the East-West Environment and Policy Institute (Hamilton 1983).

On the basis of evidence available . . . , there seems no compelling reason to move traditional shifting agriculturists out of watershed areas in the name of improved soil and water regimes if they are engaged in sustainable, stable systems. This would not be true for watersheds providing untreated water for municipal domestic use. In other watersheds, however, it might be much more effective to work with shifting agriculturists to help them in maintaining stable systems.

Given a mosaic pattern of small clearings and regrowth, many believe that the impact of extensive, sustainable shifting agriculture on changing the water table is negligible, although there seems to be no research that has tested this. While surface run-off is affected, there may be no direct effect on groundwater, springs, and wells (Hamilton 1983, 15). The same can be said for the effect of this land use on erosion and declines in productivity, except possibly over very long periods of time. On the other hand, unstable slash-and-burn activities carried out by migrants until the site is impoverished are clearly a different matter.

Converting natural forest to crop land for farming results in a number of changes. Partial or total conversion of forest to annual cropping shows increased annual streamflow yields throughout the year, but especially during the dry season. Stormflow volumes increase, and time to peak is reduced. Roads, trails, and machinery use add to stormflows and can increase local flash flooding. Any sediment washing from cropland may carry pesticides, nutrients, pathogens, heavy metals, and organic and inorganic matter. Compacted soils may lead to greater surface run-off and many rills and gullies that reduce groundwater recharge, and thereby lower water tables, reduce dry-season flow, and render wells and springs less reliable. However, lower evapotranspiration rates from annual crops, in combination with proper tillage, minimal compaction of soil, and other conservation practices, can theoretically result in an increase in groundwater levels (research in this area has been sparse). Indeed, good soil conservation practices could effectively minimize erosion, even on slopes of up to 60 percent, such as in the hydrologically benign and productive wet rice terrace fields of Bali and some other locations (Hamilton, personal communication).

Food or commodity plantations include tea, coffee, oil palm, rubber, cacao, coconuts, and bananas. The short-term hydrology and soil effects after the forest is cleared and before vegetative cover establishes itself are similar to what results from commercial wood harvesting and can include initially increased rain throughfall, a higher water table, and soil erosion. There have been few studies of soil nutrient outflow from food tree plantations. Erosion rates may increase during the period that the tree crop is being replaced with a new one; this may be negligible if the old tree is simply cut and the soil is minimally disturbed, or may be substantial if bulldozers are used. Hamilton (1983, 89–90) concludes:

> For relatively long-rotation tree crops [upwards of 15 years] in certain kinds of environments, with appropriate soil conservation practices and hydrologically sound roading, there appears to be little difference (compared to forest cover) with regard to important changes in the ... variables under consideration once the conversion has taken place.

The conditions required for this, Hamilton points out, can be extensive, and include minimal cultivation, terracing, retention of buffer strips of natural vegetation, careful roading, understory management of ground vegetation, and other special conservation measures.

Little work has been completed on the potential effects of agroforestry, but Hamilton believes that "the replacement of forest with a stable, cyclical agroforestry system is likely to have little effect on groundwater levels, streamflow timing and distribution, and sediment in streams" (Hamilton 1983, 111). Because cropping between newly planted trees is part of taungya-type systems, disturbance of the soil will occur and can result in significant amounts of erosion. Continuous cropping agroforestry systems can be sustainable if special soil conservation measures are implemented, including terracing on steep slopes, minimum tillage, and retention of streamside buffer strips. Continuous cropping systems may cover entire catchments, even though they are composed of many small parcels. Water yields will then generally be higher than that of the forests they have replaced.

Although commercial logging has many direct and indirect impacts on soil structure, effects of commercial harvesting vary considerably in response to a number of factors: the amount of canopy removed; the amount of biomass removed (including how much slash remains on the area); the extraction and removal methods; the timing with respect to wet and dry season; the soil, geologic conditions, and topography; the extent, nature, and use of roads, skid trails, and landings; the methods of slash disposal and site preparation; the promptness with which regeneration occurs (or reforestation is carried out); the presence or absence of adequate riparian buffer strips; and the nature of climatic events following disturbance (Hamilton 1983, 27). The major findings from measurement of watershed processes show that, with the exception of cloud forests, logging results in an increase in height of the water table and an increase in streamflow quantity, the effect diminishing rapidly as forest regeneration progresses. A direct cause and effect, however, between logging and flooding is not clearly established: if poor logging practices result in serious erosion, this may lead to serious sedimentation of streams that can aggravate flooding downstream, with potentially negative consequences on aquatic life, reservoir siltation rates, altered stream channels that may further increase flooding and reduce navigability, and reduced water quality for domestic and industrial use (Hamilton 1983, 49; also see chapter 8).

The removal of woody biomass can substantially deplete the nutrient bank of tropical forests (Whitmore 1984). The subsequent disruption of the tight nutrient cycle in some rain forests and the change in the microclimate and in biological activity would be even more dramatic than in the temperate zone, but Hamilton could find no studies monitoring logging effects on the nutrient input into discharging streams in the tropics (Hamilton 1983, 47). Studies in the temperate forests, such as at Coweeta and Hubbard Brook in the United States, do show substantial nutrient export (Hamilton, personal communication).

The impact of harvesting fuelwood and lopping fodder tends to decrease with distance from villages and settlements. Reduction in the forest canopy is the major cause of any hydrologic impact of sustainable fuelwood or fodder cutting, and this will result in decreased interception, increased throughfall, and decreased evapotranspiration (Hamilton 1983, 25–26). Policies with respect to use of fuelwood and fodder foliage in important watershed areas should concentrate on two factors: restricting the intensity of harvesting and controlling the way in which material is removed, to minimize adverse changes in run-off, erosion, and sedimentation (Megahan 1977; Gilmour 1977, cited in Hamilton 1983).

The effects of forest conversion to plantations depend on a number of factors, including the canopy characteristics of the stands during their life span, the type of forest or land use the plantations are replacing (wasteland, dry forest, etc.), whether the litter is collected for energy purposes (thus reducing the water absorptive capacity of the soil), and the length of the felling cycles. If cultivation is involved, measures such as contour strip planting and retention of buffer strips of native vegetation along watercourses can substantially reduce erosion and sedimentation. Once conversion from native evergreen forests to pine or broadleaved plantation species has taken place, there is little evidence of any difference in soil and hydrology variables. However, in regions with dormant seasons, streamflow was reduced substantially

when deciduous forests were replaced with evergreen conifer plantations (Swank and Miner 1968; Swank and Douglass 1974; both cited in
Hamilton 1983, 64).

CONCLUSION

The protection of key watersheds today is an investment that is at least
as sound as the dams, levees, dredging operations, fish hatcheries,
reforestation projects, water purification equipment, and other means
of replacing lost environmental services. Normally, the former will cost
but a small fraction of the large sums needed to carry out the latter,
thus saving scarce development resources from being otherwise diverted to remedial activities.

Because forests serve a variety of uses, and many people both within
and outside of the forests are dependent upon them, timber production and general forest policy should be planned within the framework
of national policy regarding overall land use. Forest policies that do not
incorporate social and environmental considerations will lead to opposition by farmers, environmentalists, and others concerned with or
dependent upon the forest, and may result in political upheaval, vio-
lence, and deliberate destruction of the resource. Forest areas that are
of critical and widespread significance, due to their rich biological
diversity or some other environmental service, need to be identified
and strictly protected or regulated.

Timber production and other alterations of tropical forests are not
necessarily incompatible with the protection of environmental services
(see also chapter 8). Significant amounts of timber and/or other products can be harvested from many forests without permanently damaging environmental services. The primary exception is biological
diversity, which can only be maintained in the long term in undisturbed forest. Most often, major impacts occur in the period between
one type of forest use and its conversion to another, when the ground is
uncovered and new vegetation has not yet established itself. Generally,
once the replacement cover has developed, these impacts diminish and
hydrological and soil characteristics become similar to that of the
original forest, although soil compaction may take decades to redress.
Even partial recovery of these services, however, can begin only if
conservation practices are implemented. Critical watersheds, such as
those with steep slopes, need to be identified and studied to determine
whether they can be managed for multiple use without compromising
the services that they provide. The continuing emphasis on fast-
growing tree plantations with short rotation cycles is cause for concern,

especially with respect to nutrient depletion. Research into this is urgently needed.

Most of the research on soil and hydrologic effects of forest alteration has been carried out in temperate regions. While basin processes are probably the same in the tropics, their different soil properties and climatic patterns warrant caution, and require considerably more research.

CHAPTER 7

The Economics of Multiple-Use Management

WHEN THEY are managed at all, tropical forests are almost always managed for timber, and other values are rarely systematically considered. However, seldom is economic justification presented for focusing exclusively on one forest use. The concept of multiple-use as applied to forests is based on the recognition that a variety of goods and services can be produced from the same land, either simultaneously or serially, and that such management can greatly increase the net value of the forest. In fact, this approach can help ensure conditions for the sustainable production of timber. The emphasis in this chapter is on multiple-use management for natural forests; the management of mixed-species timber plantations is a specialized type of multiple-use management discussed in chapter 9.*

Multiple-use management does not imply that all possible forest uses should occur in the same place at the same time. Forest management involves encouraging some uses, while discouraging conflicting ones. Determining the optimal mix of differing uses on a single forest plot requires that the costs and benefits of each be considered when designing management practices. Management problems can arise because

* For a thorough discussion of multiple-use management applied to temperate forests, see M. D. Bowes and J. V. Krutilla, *Multiple-Use Management: The Economics of Public Forestlands* (Washington: Resources for the Future, 1989). While general principles discussed there are similar, the variability of economic and political constraints facing developing countries in which tropical forests are found requires a separate explanation to enable a broader application.

multiple uses of the same forest area may also involve multiple users, multiple and conflicting management objectives, multiple time frames, and negative interactions among uses.

Figure 7.1 presents an overview of the numerous factors that must be considered when devising multiple-use management plans for natural forests. Knowledge concerning the ecology and economics of a forest is necessary, but not sufficient. Information about relevant institutions, customs, and political factors is equally important in assuring the overall success of multiple-use forest management.

There are six basic uses of forest lands:

1. timber production;
2. production of non-timber goods such as fruit, nuts, rattan, game, fish, and firewood;
3. provision of environmental/biological services such as watershed and soil protection, and conservation of genetic diversity;
4. regulation of climate and carbon sequestration;
5. recreational and aesthetic benefits, including tourism; and
6. conversion to agriculture or livestock production.

Forest conversion implies but does not necessarily entail complete destruction of the forest cover. As with any economic decision, there will be benefits and costs connected with managing a forest for either multiple uses or a single use.

Spatial relationships among uses can vary considerably. There may be different uses of the same tree. A mixture of trees and other organisms or plants within the same geographical area may serve a particular use, or uses may be separated spatially, such that different uses may be carried out at different times within a specific area (as with crop rotations).

Multiple-use systems will entail different benefits and trade-offs than single-use systems. For example, the loss of specialization in multiple-use systems can result in fewer possibilities for mechanization, lead to higher harvesting costs, and require greater management skills. Economies of scale may also be lost. Yet exploiting a larger number of spatial and intertemporal niches may result in a more efficient use of space and other scarce resources. Harvesting of different products can occur in different seasons or years, allowing more efficient use of labor. Also, production of a large variety of goods spreads the risk of crop failure. Some goods and services will be even more productive when jointly managed, such as when nitrogen-fixing *Periserianthes* (*Albizia*) *falcatara* is grown over cacao.

Figure 7.1
Tropical Forest Management for Multiple Use

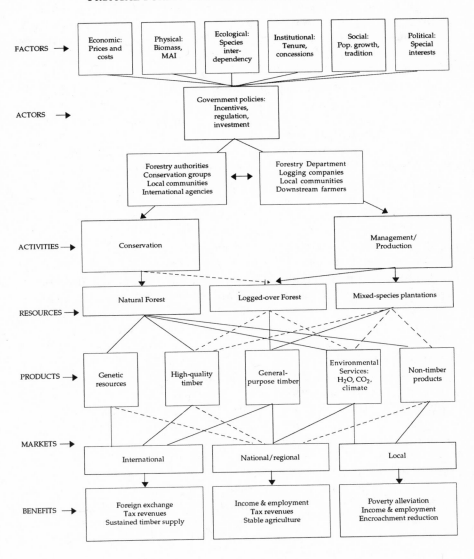

All natural forests are, to some extent, potentially multiple-use forests, as they yield a variety of products and services. The problem lies in assigning values to these potential uses in order to determine to what extent each of them should be planned for and encouraged when different uses compete with or enhance each other. There are many combinations of uses that interact in a manner similar to the combination illustrated in figure 7.2. In this example, sustained-yield timber production and recreation are combined. With increasing recreational use, timber production and its related revenues decline (figures 7.2A and 7.2B). However, the combination of revenues from both timber production and recreational use increases total revenues to some optimal level that is greater than returns from timber harvesting alone (figure 7.2C) (Hufschmidt et al. 1983), assuming that both activities occur on the same sites or adjacent sites. Attaining this optimal level of increased benefits is the goal of multiple-use management.

FOREST-RELATED INTERACTIONS

Multiple-use forestry is based on the explicit recognition and utilization of the complex ecological, economic, and social interactions associated with a forest. Only when all of these positive and negative relationships are evaluated and accounted for can the greatest value be obtained from a forest.

Many of the ecological interactions of a forest are fairly obvious. The role of forests in erosion control, soil protection, and overall water balance is now well known and widely appreciated, although the exact relationships vary and are often not clearly understood. Trees provide food and habitat for a wide variety of organisms. The regeneration and growth of a variety of plants, rattan being a good example, depend on the presence of trees to provide physical support and/or shade.

In multiple-use forests, ecological interactions can reduce the input requirements relative to those of forests managed solely for timber production or for any other single use. There may be less need for fertilizer, for example, since some species in a mixed-species forest may fix nitrogen or may rapidly cycle phosphorous. Likewise, there may be less need for pesticides in a well-managed multiple-use system, since multiple-use management may increase natural protection from pests and diseases (see chapter 9).

Planners and managers overlook these positive interactions at their peril. Indeed, ignoring the positive interactions already present in a forest may reduce economic returns from forest systems. It has been

Figure 7.2
TIMBER VS. RECREATION

A

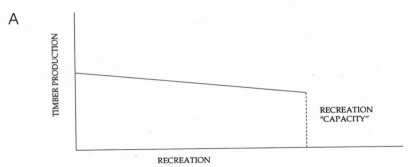

As the number of visitor-days increases, timber production falls.

B

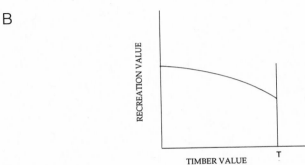

As the annual value of timber increases, the annual value of recreation declines.

C

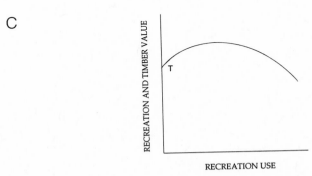

Recreation value and timber value may be maximized at some optimal number of annual visitor days for recreation.

BASED ON: Hufschmidt et al. 1983.

estimated that in the Amazon basin, for example, some 200 tree and fish species are interdependent to some degree (Smith 1981). The fish feed on fruit or nuts from the trees. In turn, the action of the fish in physically breaking open the nuts helps the seeds to germinate; the fish also help disseminate the seeds from the fruit. The potential harvest of fish from this region is estimated at 250,000 tons/year. Fish species that share this ecological relationship with trees account for about three-quarters of the total catch. However, when considering the conversion of these forests to rice fields and buffalo ranches, the potential loss from reduced fish harvest is not seriously evaluated. This is true even in cases when it is not clear if, in strictly financial terms, the value of the fish is outweighed by the value of rice and water buffalo (Myers 1984).

The social interdependence between people and forests is now commonly recognized. Animosity, encroachment, and sabotage by local communities who receive little or no benefits from timber production may doom timber-producing forests and plantations, especially when traditional uses of the forests are thereby precluded or prohibited. In many countries, environmental groups perceive timber harvesting as detrimental to the environmental functions of the forest and are attempting to increase their political influence on policy decisions.

Such arguments underscore the economic interdependence among various functions of the forest as well; economic interdependence is a significant aspect of multiple-use management, and by definition has several positive implications. Perhaps the most important is risk reduction. Having more than one source of revenue in a given area reduces the overall level of economic risk. When one species or crop fails, others may continue to provide revenue.

Another important implication of economic interdependence arises in reference to the long period between harvests of timber from the same stand. The waiting period required for returns to investment in timber may favor interim dependencies on other forest products, especially for smallholders who may have some difficulty obtaining credit. If there were perfect capital markets, then waiting 30, 50, or even 70 years to receive the returns on an investment would present no problem because money could be borrowed against future revenues. With imperfect capital markets, however, little or no money may be available for borrowing on these terms. Even when future returns can serve as collateral for borrowing, it may be desirable to derive other income from the land in order to generate short-term cash flows. It is clearly advantageous for project managers to be able to include additional income-producing activities (fuelwood, fodder, fruit and nut production, etc.) during the extended periods necessary for hardwood timber harvesting. Such strategies would be financially justified if the inclusion of these non-timber goods and services had positive or no effect on timber production.

Even when the inclusion of non-timber goods and services will decrease the overall financial net present value of the forest, the inclusion of these activities may still be a preferred option if they increase overall utility by providing annual revenue flows. By improving the cash-flow profile, the project may become more attractive to investors who might otherwise be unwilling to invest in a project involving only long-term returns.

OVERVIEW OF MULTIPLE-USE ECONOMICS

Multiple-use management is used to maximize the net present social value of forest land. "Social" value is used here to include all the values of interest to society, whether or not their costs and benefits are measurable in the marketplace. Product valuation within multiple-use forest management is difficult even in developed countries, with readily available data on market prices and the attributes of the forestland in question (Bowes and Krutilla 1989, 88); valuation in the context of tropical forests, given the increased complexity of known and unknown forest goods and services, presents an even more formidable task.

Multiple-use economics is used to determine (1) whether to employ multiple-use management and (2) the extent to which it is appropriate to do so. The optimal combination of uses is the one that maximizes the net present value of all uses combined. The analysis should include all relevant social costs and benefits, including the effects of interactions among uses and other externalities. Additional uses should be pursued if their social benefits exceed social costs. The wide array of inputs and conditions required to minimize costs or to maximize volume or value of five different forest uses is illustrated in table 7.1.

Dominant-use management is a relatively simple version of multiple-use management. It involves two basic steps. In the first step, the single use that maximizes the net present social value of the land is selected as the dominant use. Theoretically, the classification of land for the dominant use could include ecological, economic, political, or social considerations. In practice, the difficulties of analysis based on these criteria lead many countries to develop land-use classification schemes largely on the basis of easily measured physical characteristics of the land, such as climate, soil, gradient, and hydrology, without direct reference to dominant-use management per se.

Table 7.1
TROPICAL FOREST MANAGEMENT FOR MULTIPLE USE: OPTIMIZATION

Type of Output	Maximum Volume (1)	Maximum Value (2)	Minimum Cost (3)	Maximum Net Value (4) = (1,2,3)
TIMBER PRODUCTION (including local consumption & exports)	• Standing volume • Growth • MSY • Preference for fast-growing species • Conversion into plantations	• Timber prices by species • Balancing of volume of general-purpose timber against value of high-quality timber	• Management costs • Planting costs • Logging costs • Transportation costs • Opportunity cost of land • Interest rate • Preference for more accessible natural forest	• Volume & growth • Prices by species • Costs • Optimum rotation • Preference for high-stumpage timber
FUELWOOD PRODUCTION (including charcoal)	• Biomass & growth of woody species • MSY	• Price of fuelwood • Price of charcoal • Price of substitutes	• Management costs • Harvest & transportation costs • Opportunity cost of land & trees	• Prices (& yields) • Costs • Net value

NON-WOOD PRODUCTION (including food, fodder, fiber, & medicine)	• Density and productivity of species producing non-timber goods of commercial value	• Price by product • Balancing of high productivity against high-value products • Price of substitutes	• Management costs • Harvesting costs • Transportation costs • Opportunity cost of trees • Opportunity cost of land	• Prices (& yields) • Cost • Net value
SOIL & WATER CONSERVATION (including climate)	• Continuous ground cover	• Scarcity value of water of given quality • Social value of flood control • Shadow price of soil	• Management costs (including monitoring & protection) • Value of other forest uses forgone • Opportunity cost of land	• Values (& Q's) • Costs • Net value

GENETIC RESOURCE CONSERVATION (including wildlife & wilderness)	• Survival of the largest possible number of species (both plants and animals)	• Potential commercial value
		• Scientific & educational value
		• Amenity value
		• Existence value
		• Option value
		• Irreversibility
			• Management costs (including protection, monitoring and research costs)
			• Opportunity costs
			  –of forest
			  –of land
				• Market values
				• Non-market values
				• Costs
				• Net social value

Once the dominant use of the land has been assigned, the next step is to determine the extent to which other uses, if any, should be allowed. This decision is resolved by examining the nature of the interactions among uses. The ecological relationships among different uses are extremely complex, as are the social and economic interactions to be considered. The important aspect of these relationships for forest management is the degree to which each use will diminish or increase the productivity of the dominant use. The ecological interactions among uses must first be examined, since, if their net effects are not positive, there is no point in analyzing economic and social interactions. There are four possible ecological interactions:

1. The interactions may be completely negative; in this case, the dominant use should be chosen as the sole use, since nothing would be gained and much lost by combining it with other uses.

2. There may be only positive interactions among uses, in which case the choice is also clear. The additional use(s) should be included, since this would increase the net present value of the land.

3. There may be no physical interactions among uses, and no competition for space, light, or other limited factors. In this case, inclusion of additional uses is justified on a case-by-case basis. For example, rattan and timber might be intermingled without one type significantly affecting the growth of the other. However, one of these uses may be disqualified if there are negative social or economic interactions. There may be increased harvesting costs, decreased economies of scale, or other increases in the cost of producing the dominant product. The mere physical presence of one use may thus cause negative economic interactions between uses that must be included in determining whether and to what extent two or more uses should be combined or left separate.

4. There are both positive and negative interactions among uses. This is the most interesting and by far the most common case. To determine whether additional uses are justified and to what extent, the net present value of different combinations of uses must be quantified and compared. If the net interdependence is negative with all possible combinations, the indicated choice is to manage only for the single, dominant use. If the net effect is positive, the additional use should be included and expanded up to the point where its marginal contribution to the net present value of the land becomes zero.

Cost-benefit analysis is an economic tool used to define and compare the net present values of alternative courses of action, thereby enabling a manager to arrive at the most efficient allocation of resources. This is accomplished by attaching values to all relevant variables—timber and non-timber goods, environmental services, labor costs, social benefits and costs—whether or not they are priced in a market. Any benefits or costs that will be incurred in the future are discounted to the present in order to obtain comparable values.

COST-BENEFIT ANALYSIS

The difficulties involved in identifying costs and benefits and in choosing an appropriate discount rate make it both conceptually and empirically difficult to determine the "true" social value of management alternatives. Further, the value of forest benefits will be viewed differently by different users and beneficiaries. Discrepancies between who bears the costs and who receives the benefits also pose barriers to ensuring the most efficient allocation of uses. All of these thorny issues are not arguments against the use of cost-benefit analysis; rather, they emphasize how easy it is to ignore or undervalue many social, environmental, and economic effects of decisions. At the very least, cost-benefit analysis helps to make priorities and values explicit, thus ensuring that the trade-offs involved in decision making are consciously understood.

The most efficient allocation of resources is that combination which maximizes net present value. This is found by computing the annual stream of benefits and costs associated with each mix of activities. The discount rate is then applied to convert future benefits and costs into present values. However, each element of this procedure presents problems that make the analysis quite complex.

Valuation of Benefits and Costs of Forest Goods and Services

Assigning value to forest goods and services is often quite difficult. Although the importance of non-timber forest products and services is increasingly being recognized, information about their quantity (supply) and value (price) is often inadequate or lacking. Many non-timber products play a significant role in local consumption. Some products such as fuelwood and poles are collected and consumed directly by households and are also traded on local markets. Thus, for certain forest products—fuelwood, fish, fruit, etc.—at least some approximation of value can be made on the basis of market prices. If there are market imperfections, however, market prices may not reflect the true social value of these products and the opportunity cost of resources being used. In such cases, adjustments or corrections of market prices for distortions, known as shadow pricing, is necessary.

In addition to timber and non-timber products, many forests, especially natural forests, provide many indirect services such as soil or watershed protection, habitats for wildlife, conservation of biological diversity, carbon sequestration, and opportunities for recreation and tourism. These services must be taken into account in economic analysis, but because they are not traded in markets, assigning values to them becomes a far more difficult task than the valuation of non-timber forest products.

While the determination of the value of non-market goods and services is fraught with difficulties, several methods have been devised for arriving at evaluations that may be sufficiently realistic for many economic decisions. Shadow pricing is the process of deriving prices for a good or service when there is no monetized market or when the market fails to price goods based on their true value. The following three relationships between market and shadow prices are possible (Gregerson 1982): (1) the market price exists and reflects willingness to pay (w.t.p.), in which case market prices and shadow prices are the same; (2) market prices exist but due to market imperfections or policy distortions they do not reflect willingness to pay and therefore shadow pricing is necessary; and (3) no market price exists, but market prices for similar goods and services might help in determining shadow prices.

Numerous methods have been developed to determine shadow prices. Several of these valuation methods have been based on three major data sources (Sfeir-Younis 1987): (1) observed economic behavior, (2) surrogate market values, and (3) elicited responses (see appendix 7.1 for more on all three).

Whatever the merits of the various approaches, it is equally important to understand the shortcomings of such methods. Areas where additional work is needed include the problem of using w.t.p. measures when income is very unevenly distributed, when human lives or aesthetic benefits are being measured, and when biological diversity and irreversible change are being valued.

Boundaries of Economic Appraisal

Another particularly difficult problem arises from the enormous differences in the way costs and benefits are valued by private firms and by society as a whole. Benefits in the form of profits from the sale of timber, fuelwood, or other marketable products will accrue to the private owner or concessionaire. These financial benefits provide the only incentive for private investment. Other benefits such as watershed protection may be critical for enhancing social welfare, but there is no incentive for including these considerations in private decision making. Similarly, the only costs incurred by private owners are those that will affect the profit to be made from selling a forest product. The public, on the other hand, may well consider the costs of downstream siltation, loss of soil fertility, or other similar effects to be very high, often even higher than the value of timber.

Because positive and negative externalities do not accrue to the private forest owner/manager/concessionaire, there is little or no incentive to include these externalities in private management decisions. Further, the lack of a market for many non-timber goods and services and hence the absence of a mechanism for readily pricing them leads to the misallocation of forest goods and services as well as inputs such as labor and capital. From society's point of view, the most efficient allocation of resources can only be determined by including all inputs, outputs, and externalities relevant to an economic decision, regardless of whether or not they are priced in the market.

In some cases, society as a whole and private firms may view costs and benefits in completely opposite terms. For example, a private firm may find it advantageous to increase mechanization in order to decrease labor costs, thereby providing a net benefit for the firm. However, in the presence of high unemployment, the loss of jobs may be considered a net social cost. In addition, mechanized harvesting may damage soil and increase erosion, reducing the value of other forest goods and services, as well as hampering forest regeneration. If the timber producer does not own the land or use the services, there will be no incentive to mitigate these effects. Yet all these effects impose a real cost on society.

The significance of each type of forest use depends on both the intensity of the use and the extent of any related externalities. This will cause the significance of different uses to vary according to the interests of those consulted, as summarized in table 7.2. For example, timber is mainly sold in national and international markets, while non-timber products (including fuelwood) are sold and consumed primarily in local markets. Local communities may therefore be more interested in having access to fuelwood, poles, and food from the forest than in sizable timber harvests. Conversely, local communities may view forest preservation in and of itself as of little or no importance. The international community, on the other hand, may wish to save forests for the sake of preserving biological diversity and other benefits. These varying levels of significance reflect who is most directly concerned with particular forest uses, and who is therefore most willing to pay for or otherwise promote particular forest uses.

Table 7.2
TROPICAL FOREST MANAGEMENT FOR MULTIPLE USE:
PRODUCTS AND LEVELS OF SIGNIFICANCE

Type of Output	Local Level	National Level	Regional Level	International Level
Timber Production	Minor to Moderate	Major	Moderate to Major	Major
Fuelwood Production	Major	Moderate	Minor	Insignificant
Nonwood Production	Major	Moderate	Minor to Moderate	Minor
Soil & Water Conservation	Moderate	Major	Moderate to Major	Minor
Genetic Resource Conservation	Insignificant	Minor	Moderate	Major

Local: fuelwood and non-wood products
National: timber and watersheds
International: timber and genetic diversity

The Discount Rate

The discount rate represents the implied time preference held by an individual, group, or by society as a whole. It is used to compare diverse projects by providing a common currency with which to compare the total flow of costs and benefits for each project. This is accomplished by discounting benefits and costs into present-value terms.

Which discount rate to use for a particular analysis can be a matter of considerable controversy. In general, private firms will use higher discount rates than society as a whole. This is because private firms are usually interested in avoiding the risk of waiting to receive benefits. Society as a whole, however, may be more willing to spread out benefits over time, in spite of the assumption that doing so entails greater risks. Also, society may be in a better position to pool risks from a large number of independent projects. Thus, when both private firms and governments invest in forest products, they may use a different discount rate for financial analysis.

To add to the complications, poor rural communities will often have a very high discount rate. The immediate pressures of hunger and poverty force people to forgo possible future benefits from a forest in order to provide goods they need for immediate survival, such as food, fuelwood, and fodder.

MANAGEMENT WITH TIMBER AS THE SINGLE DOMINANT USE

A particular optimum harvest time is often dictated by the duration of concession agreements. When this is the case, the value of the timber at harvest time as well as the sum of maintenance, harvesting, and replanting costs are discounted to derive the net present value of the investment.

When a rotation period is not assigned, the optimal cutting time can be derived by the following method. It should be noted that this formula specifically applies to a managed forest, in which all trees are the same age and are intended for timber production. The formula is based on the assumption that future stumpage values can be accurately calculated on the basis of the growth rate (specific for each tree stand) and the net timber price. While these two caveats limit the utility of the analysis for individual cases, the formula still remains useful in demonstrating the general effects of various government policies and management strategies on the optimum cutting time.

The optimum cutting time will yield the highest net present value of the land. Therefore, trees should be harvested when the marginal cost of not harvesting (allowing the trees to grow one more year) equals the marginal benefit of harvesting (receiving the revenue for the timber). At that time, the following condition will be met:

$$(PV)'_{t*} = rPV_{t*} + rW_{t*}$$
$$\text{or}$$
$$S'_{t*} = rS_{t*} + rW_{t*}$$

Where:

P = net price of timber (market price − operating cost)
V = volume of standing stock (as a function of growth rate)
$(PV)'$ = the incremental change in stumpage value (PV) from one year to the next
$t*$ = optimum cutting time
r = the discount rate
W = capitalized value of the land
$S = PV$ = stumpage price
S' = incremental change in stumpage price from year to year

At time t*, the annual increase in stumpage value will equal the opportunity cost of capital tied up in the trees plus the opportunity cost of the land, where "opportunity cost" refers to the forgone benefit from the next best alternative.

The economically optimal rotation period occurs before trees have reached their maximum stumpage value, because of a positive discount rate. If the discount rate were to increase, the optimal cutting time would decrease. This is because both the opportunity cost of land and the opportunity cost of capital are directly related to the discount rate and will therefore increase, thus increasing the marginal costs of waiting to harvest. A shorter rotation period will lower these costs. Similarly, if the private discount rate is higher than the social discount rate (as is usually the case), then the optimum rotation period for private owners will be shorter than the socially optimal one.

Taxation influences the optimal cutting time in a number of ways, depending on the type of tax. A tax on the value of standing timber would shorten the rotation, a tax on timber harvested would lengthen the rotation, while a tax on profit would leave the rotation unaffected.

An increase in the opportunity cost of land would shorten the optimal rotation, since the land could more profitably be used for some other activity. A striking example of this is found in Brazil, where government-sponsored incentives to raise cattle or to otherwise clear the land have increased the opportunity cost of forest land. There is thus little incentive for landowners to invest in forest production and management when they can earn higher returns from their land by converting it to other uses (Schmink 1986).

MULTIPLE-USE MANAGEMENT FOR TIMBER WITH OTHER FOREST USES

When non-timber forest uses are combined with timber, the same basic equation can be used with the addition of non-timber benefits:

$$S'_{t*} + B'_{t*} = rS_{t*} + rW_{t*}$$

Again, the timber should be harvested when the marginal benefits of delaying harvest equal the opportunity costs of delaying harvest. The benefits of delay include the annual increase in value from timber growth, S', plus the flow of benefits from non-timber uses, B'. The costs include the income forgone by delaying harvest revenues plus the costs of delaying benefits from future harvest cycles. When additional uses are not traded on the market, economic efficiency requires that the increase in multiple, nonmarketed benefits must exceed their own direct costs plus the opportunity costs they impose in the form of reduced production for commercial uses.

How will the addition of other uses affect the optimal rotation period for timber? This would depend on the nature of the added uses. Because of the varied nature of possible benefits, there is often no direct correlation between the extent of benefits and the age of the standing stock.

Since the value of additional goods and services, such as soil and watershed protection, diminishes with the cutting of timber, it becomes economically efficient to keep the timber standing for longer periods of time. Efficient management with certain other uses such as fuelwood collection may only require that particular cutting practices are followed and will not depend on the length of the industrial timber cutting cycle. When non-timber benefits (B) are sufficiently large, there may be no age at which the marginal benefits of cutting are as high as the marginal costs. In this case, it would never be economically efficient to cut the trees. The underlying assumption for these relationships is that the stand will consist of trees that are the same species and age; in a multiple-use management scheme, trees would likely be of variable ages and species, increasing the potential value of B commensurately. Under this alternative assumption, if (rS_{t*}) and (rW_{t*}) can be valued, the likelihood that total conservation would prove economically efficient is even higher.

MULTIPLE-USE MANAGEMENT WITH INSUFFICIENT INFORMATION

In order to apply multiple-use theory, data is needed on the full costs and benefits of marketed and non-marketed forest goods and services. Such data are usually unavailable. In addition, information is required about the types and extent of interactions among uses in the same forest.

In the absence of such information, certain guidelines may nevertheless be suggested. "Back-of-envelope" estimations or heuristic guidelines can be developed based on perceived costs and benefits. Simple sensitivity tests can suggest whether changes in given values will have much effect on rotation cycles and other management practices. Timber harvesting generally causes damage—at least in the short run—to non-timber goods and environmental services. If simple conservation practices have low costs and potentially large benefits, they should be regularly incorporated into management. When timber harvesting or other intensive uses of forest may result in irreversible consequences (destruction of genetic resources, loss of watershed protection, etc.) such uses should be postponed until further information is acquired.

The compatibility matrix presented in table 7.3 presents a number of interactive effects among major types of forest uses. Management practices or restrictions are illustrated which will increase compatibility, and hence, the value of the land.

The extent of compatibility in table 7.3 varies from "generally compatible" to "generally incompatible." As mentioned previously, the dominant use or primary forest output should first be determined, then the effects of adding various secondary outputs and uses assessed. Thus, if timber production is the primary output, there are trade-offs or costs involved with managing for other uses of the land. If the forest is also used for fuelwood production, for example, there are enforcement costs of preventing encroachment from neighboring communities. Standing timber, especially young growing stock, could be damaged if fuelwood collectors are indifferent to the type of wood they use, or simply as a result of their gathering activities. On the other hand, gathering undergrowth and brush for fuelwood can, if managed sustainably, reduce competition, encouraging greater regeneration of timber species. If there is a charge for fuelwood gathering, this will provide some short-term income to the owner, thereby enhancing investment in the land.

Timber production may also be successfully combined with soil and water conservation when techniques such as selective logging, which causes minimal disruption of forest undergrowth and soil, are employed. Sustainable timber management techniques will necessarily involve the conservation of soil and water, at least to the extent necessary to support long-term timber production.

Genetic resource conservation cannot be combined successfully with timber production within the same forest compartment unless particular genetic pools are being conserved or expensive logging methods are used (see chapter 10). Generalized species that prefer to dwell in and around open spaces will either not be negatively affected by timber harvesting or may even be enhanced. Logging methods may be chosen that leave corridors of intact forest. Certain (though probably never all) species may thus be preserved within these corridors. It may also be economically viable to use very expensive logging methods to extract extremely valuable wood (such as mahogany). The preferred means to reconcile conservation of genetic resources and other biodiversity is to select representative compartments for total preservation within a matrix, preferably extensions, of production forest (see chapter 10).

If something other than timber production is the primary use, then compatibility between the two uses can change. For example, if genetic resource conservation is the primary output, timber production generally becomes incompatible.

INFORMATION NEEDED FOR BETTER MULTIPLE-USE MANAGEMENT

Considering the qualitative trade-offs described above could help improve management systems. However, determination of the most efficient allocation of uses also requires quantitative data about the value of non-marketed goods and services and about the interactions between uses in the same forest area.

No such studies exist at this time, but two related case studies may be mentioned. In the first study (Peters et al. 1989; Multiple-Use Case Study 1 in appendix 7.2), research was conducted which produced an extensive inventory of a single hectare of Peruvian forest. The net present value of various forest goods (fruit, latex, and timber) was calculated. Under two different conditions for selective cutting, fruit and latex were found to account for up to 98 percent of the total net present value of the forest.

In the second study (Anderson 1987; Multiple-Use Case Study 2 in appendix 7.2), trees were added to degraded land in Nigeria on which cattle and farm crops are also produced. The trees not only protected the land from soil erosion but also increased soil fertility. Farm productivity increased and there was additional income from wood and fruit production from the trees. The internal rate of return for the project increased from about 5 percent up to 14 to 16 percent.

These two case studies indicate that financial incentives for forest preservation or afforestation can increase significantly if non-timber goods and services are identified and their value measured. If the full economic valuation of the forest is calculated properly, it can be used in developing guidelines for public management policy and for private investment.

INCENTIVES FOR MULTIPLE-USE PRACTICES

As noted at the beginning of this chapter, multiple-use forest management is rarely practiced nowadays in the humid tropics, even where it is economically justified. Notable exceptions include home gardens common in Southeast Asia, to be described in chapter 9. In these cases, it is the forest owner who benefits from multiple-use management, since

Table 7.3

TROPICAL FOREST MANAGEMENT FOR MULTIPLE USE: COMPATIBILITY MATRIX

Primary Output ↓ / Secondary Output →	Timber Production	Fuelwood Production	Non-wood Production	Soil and Water Conservation	Genetic Resource Conservation
TIMBER PRODUCTION	• Trade-offs between high-quality & general purpose timber • Complementarity between high-quality & general purpose timber (shade, pollination)	• Encroachment • Possible damage to standing timber • Thinning • Short-term income • Trade-offs but generally compatible	• Encroachment • Possible damage to standing timber • Short-term income • Biological interdependence • Trade-offs, but generally compatible	• Imposes constraints on logging methods • Complementary with sustainable yield • Management and selective or strip logging	• Generally incompatible except for: a) Generalized species b) Corridors c) Very select harvest of high value species (by nondestructive harvest methods)
FUELWOOD PRODUCTION	• Selective logging of high-value species not incompatible • Additional income • Logging may provide more profit	• Fuelwood from natural forests vs. fast-growing plantations—village forest • Fuelwood vs. charcoal vs. substitution	• Generally compatible in natural forest and multi-species plantations • Additional income • Constraint on fuelwood species	• Imposes constraints on access and harvest • Complementary with sustainable yield • Additional value • Trade-offs, but compatible	• Generally incompatible except for managed fuelwood species • Corridors for general species

NON-WOOD PRODUCTION	• Compatible except where logging or silvicultural methods damage non-wood production • Additional income	• Compatible for only certain species • Encroachment • Additional income • Possible damage to non-wood production	• Choice of areas • Choice of species • Densities • Local vs. national and export market	• Generally compatible • Additional value • Constraints on fodder harvest • Ground cover	• Imposes constraints on the collection of plants & animals • higher management costs • Trade-offs, but compatible
SOIL & WATER CONSERVATION	• Compatible only if logging methods do not disturb, compact, or expose the soil to erosion • Higher logging & management cost but additional value	• Encroachment • Possible damage to ground cover and the soil • Additional value • Compatible if properly managed	• Encroachment • Possible damage to ground cover and soil • Additional income • Compatible if properly managed	• Selection of critical watershed areas, densities, and cover species • Monitoring and protection • Substitutes • Rehabilitation of degraded areas	• Generally compatible but most watersheds degraded or poor in species • Additional value
GENETIC RESOURCE CONSERVATION	• Generally incompatible	• Generally incompatible	• Generally incompatible except for collection of species samples for research	• Generally compatible • Watershed areas can also serve as corridors	• Choice of areas • Choice of species • Generalized vs. specialized species

many of the forest products are either directly consumed or sold by the owner.

Forest services from which the owner derives few or no benefits are another matter. Unless the benefits of multiple-use management can be captured by the owner, there may be no incentive to practice this system. It could be financially disastrous for a timber owner not to harvest timber at the optimal cutting time in order to provide environmental services such as watershed protection to nearby communities.

This common mismatch between those who pay and those who benefit from multiple-use forestry management has prevented these practices from being more widely adopted. Because these environmental benefits and costs are not usually traded in the marketplace, there is market failure, and society is justified in assigning ownership of its forest lands based on their dominant use. What benefit does a government provide by holding forest lands that it cannot adequately police in situations where it cannot derive the full value of the land?

Three categories of insecurely held or state-owned land can be defined. The first is forest lands that involve only local externalities. These lands could be privately owned and managed, since the benefits would accrue to the owner. The second is forests that provide community-wide externalities such as fish-spawning areas or local watershed protection. Even if there were no significant externalities, economies of scale might justify community ownership. The third is forests that provide far-ranging externalities of regional, national, or international importance, such as large watersheds or critical habitats. These forests should be managed with the interests of society as a first priority. Only in this way can the incentives for obtaining maximum value from the land be fully internalized.

Taxes and subsidies can promote multiple-use management on privately held lands by providing incentives for the utilization of nonmarketed goods and services. Taxes could be imposed to discourage practices such as clear-cutting, which contribute to the destruction of watershed or soil erosion. Subsidies could be used to promote more beneficial practices such as the use of selective cutting techniques to maintain biological diversity.

CONCLUSION

There is evidence that the case for multiple-use management can often be based on financial analysis alone, without resorting to shadow pricing to correct market failures or distortions. The dynamics of multiple-use management make such projects attractive investments even if the optimal harvesting age for timber is increased as a result. Increased cash flow, better relations with nearby communities (thereby decreasing the likelihood of encroachment and of opposition from environmental and indigenous rights groups), increased productivity per hectare, and other benefits may all be captured by the owner(s) or by society as a whole. For these reasons, multiple-use management will help ensure the long-term sustainability of production of timber and other tropical forest goods and services.

APPENDIX 7.1

Methods for Assigning Values to Non-Marketed Goods

VALUATION METHODS BASED ON OBSERVED ECONOMIC BEHAVIOR

These methods assess environmental services by determining how changes in them are related to the supply or productivity of goods and services traded in the market. The approaches most frequently used include:

Changes-in-Productivity Approach. This approach measures the extent to which changes in environmental attributes (e.g., pollution, erosion) affect the productivity of different factors. Productivity of goods and services such as land, labor, fertilizer, farm machinery, or of particular resources such as forests or fisheries, may be examined.

Loss-of-Earning Expenditure Approach. This approach values changes in labor earnings (income) due to the effects of environmental degradation. Pollution, for example, often results in increased medical expenses. These expenses can be used as a proxy for benefits when they could be avoided if there were less environmental degradation.

Cost-Effectiveness Approach. This approach is used to decide the least-cost method for environmental improvement. It is useful when benefits

This section borrows extensively from Sfeir-Younis 1987. Other highly recommended sources include Vincent et al. 1991 and Dixon and Hufschmidt 1986.

are difficult to estimate in practice, when most alternatives will result in similar levels of benefits, and when goals or standards are fixed and agreed upon by policy makers. The analyst looks at the least-cost means of achieving those goals.

There are several extensions of these methods, the most useful ones being the replacement-cost approach, the compensation approach, and the wage-differential approach.

Replacement-Cost Approach. Costs which society must incur to replace assets that have been damaged or depleted are calculated. The primary assumptions used in this approach are: (1) the real value of damages can be accurately measured; (2) the irreversible loss of an environmental asset can be replaced by another asset of equal value to society (over space and time); and (3) there are no externalities associated with the necessary expenditures. The estimated replacement cost is used to approximate the benefits gained from avoiding the damage that is occurring or which will occur in the future. The calculation for this proxy is "lower-bound," and often does not accurately measure the true value of benefits of a given environmental intervention (e.g., protection). The approach is also very limited when dealing with assets that do not have perfect substitutes.

Compensation Approach. The analyst tries to estimate the cost of relocation of physical assets or of individuals due to environmental degradation or other changes in the environment. This approach is useful in cases where an asset such as a factory needs to be relocated to avoid damages to another environmental good (e.g., industries polluting rivers) or when it is necessary to compensate people, for example, due to the construction of dams. The determination of an appropriate formula for compensation is a key issue in the application of this approach. In practice, estimating the cost of relocation of physical assets is simpler than calculating the cost of compensating individuals.

Wage-Differential Approach. The analyst deals primarily with changes in wages that are due to changes in environmental attributes associated with a particular job. The application of this approach assumes that labor markets are competitive, that the wage rate is equal to the marginal productivity of labor, and that the supply of labor varies according to the attributes of any given job. If these assumptions are valid, then higher risks and levels of pollution in equivalent jobs will be reflected in higher wage rates, since higher wages will compensate for the lower environmental quality. Wage differentials can therefore be used as proxies for estimating benefits from environmental improvements.

Other cost-related approaches, such as the mitigation-cost and prevention-cost approaches, may also prove useful in analyzing certain environmental decisions. In these, the analyst focuses on the cost that is necessary to reduce or prevent existing environmental damages. These approaches often provide a lower bound in estimating total benefits.

VALUATION BASED ON SURROGATE VALUES

These methods provide estimates of implicit values based on prices paid for other marketed goods. Three approaches that utilize surrogate prices have proven useful: the property-value approach, the travel-cost approach, and the shadow-project approach.

Property-Value Approach. This approach estimates changes in land values as a function of several parameters, including changes in environmental attributes (e.g., soil erosion, pollution, waterlogging). The approach assumes that land values reflect future income to be generated from the land, that changes in the quality of land are visible, and that land prices will therefore vary according to changes in land quality. These assumptions are sometimes unrealistic, since attributes such as erosion can be difficult to see and measure. Moreover, this approach has some limitations in the context of developing countries because land markets are often inactive or imperfect. Rural land values in particular often reflect values that are not related to productivity. For example, land is purchased for security reasons, for land speculation purposes, or for increased prestige. Such imperfections often cause analysts to use income forgone rather than revealed market values as the basis of analysis. This is done by estimating the value of land as the discounted value of future income streams "with" and "without" a change in the environment. This is a "flow" value and tends to underestimate the true value of land to society. In countries where land is a major constraint to development (or where each forest contains unique biodiversity), the reduction in size of the country's stock of forest land will put a high premium on each remaining unit of productive land. Thus, irreversibility will also affect the land values; this is not accounted for in any of the above-mentioned approaches.

Travel-Cost Approach. This approach estimates the willingness to pay for using a particular resource as a function of expenditures incurred to use it. The method has often been applied to evaluate the benefits from recreation. The demand for the resource (e.g., the recreational site) is derived from a relationship that takes into account the time and monetary costs of traveling to the site, and various socioeconomic variables. An inverse relationship between frequency of use and distance is expected. In estimating this demand curve, several assumptions are made: similar preference functions of individuals in the same zone or distance, and their equal reaction to increased value of travel and to changes in access fees to the resource.

The principal limitations of this approach are that the total value to estimated benefits reflects only the willingness to pay of those who actually use the facility or the environmental resource and not the benefits to society. Benefits are often underestimated when population density is very high near the area under consideration. Since this approach could be refined to reflect the value of time spent in traveling, this approach could be useful in evaluating benefits from some social forestry products (i.e., valuing time spent in gathering fuelwood) and protection of national parks.

Shadow-Project Approach. The analyst attempts to estimate benefits received (or forgone) by looking at alternative ways of providing an environmental service via the marketplace. Thus people may benefit today, for example, from the use of clean beaches and lakes or from national parks, but in their absence, society would offer such alternatives as swimming pools or public parks. Since the value of providing these marketed goods can be easily obtained, the analyst would estimate the benefits of environmental services not traded in the market by observing the value of the marketed private good. The principal assumptions are that perfect substitution exists between the market good and the non-market good, that the proposed market good will provide all the services supplied by the non-market good, that the original level of supply of services is desirable, and that the overall cost of the shadow project does not exceed the total value of the "cost" service provided by the natural environment. This approach could be usefully applied in decisions dealing with drinking- or irrigation-water supplies, alternative sources of energy, and the like.

VALUATION BASED ON ELICITED RESPONSES

These valuation methods are used to determine the value of benefits based on data directly elicited from different users of a given resource.

These non-market data are obtained through surveys, questionnaires, bidding games, and voting. Collectively, they are termed "contingent valuation" (Mitchell and Carson 1989). These methods involve gathering data on such matters as individuals' willingness to pay for higher-quality water or air, or for retaining the future productive capacity of a resource. Examples of applications in developing countries of contingent valuation methods include Briscoe et al. 1990; Whittington et al. 1989; and Whittington et al. 1991 (draft).

Other methods of valuation have been developed to handle multiple objectives. Examples of such objectives are "safe minimum standard," achieving "intergenerational equity," and preserving "genetic diversity." Moreover, there are various methods that are useful in assessing macro and regional trade-offs, which would enable analysts to incorporate environmental concerns into macroeconomic policy. The most well-known methods are input/output analysis, trade and investment models, and material balance models.

APPENDIX 7.2

Case Studies of Multiple-Use Forestry

MULTIPLE-USE management of natural forests, which incorporates the production of non-timber goods and services into timber production, is rarely studied because of the difficulty of identifying and measuring non-market commodities and services provided by the natural forests. Some studies have addressed these issues, however, in the context of forest plantation projects in developing countries. Two case studies demonstrate the significance of non-timber goods and services and how to manage multiple-use forests.

The first is a study of the significant economic value of non-timber goods in the natural tropical forest of Peruvian Amazonia (Peters et al. 1989; see also chapter 5, *Non-Timber Products Valued for Reasons of Tradition or Uniqueness*). The other study deals with attempts to increase the net economic benefit of farm production by the use of afforestation techniques in the arid region of northern Nigeria (Anderson 1987).

CASE 1. VALUATION OF TROPICAL FOREST IN PERUVIAN AMAZONIA

The purpose of this case study was to calculate the net present value of the forest products in a 1.0-hectare stand of the Mishana forest on the Rio Nanay 30 km southwest of Iquitos, Peru, in order to justify the conservation and rational use of the forest and to maximize its economic benefits.

A systematic inventory of the single hectare of forest showed 50 families, 275 species, and 842 individual trees of greater than 10 cm in diameter. Of these, 72 species (26.2 percent) and 350 individual trees (41.6 percent) yielded products that have an actual market value in Iquitos.

Edible fruits were produced by 7 tree species and 4 palm species, 60 species were commercial timber trees, and 2 species produced rubber.

The value of the forest resources, including fruit, timber, and rubber, was assessed at the actual market value. The yield of useful products per unit of time was determined for each resource. The net revenues generated by the sale of each resource were calculated based on current market values and the costs associated with harvest and transportation. Two different harvest scenarios were used. The first involved the selective removal of all existing timber greater than 30 cm in diameter in year 0, year 20, and year 40, with a final cut of all remaining trees (projected to have a minimum diameter of 30 cm) in year 65. Annual collection of fruit and latex was assumed throughout the 65-year cutting cycle. The second scenario, that of sustainable yield, assumes selective timber removal (30 cubic meters/harvest) on a 20-year cutting cycle with annual fruit and latex collections in perpetuity.

Using the criteria for the first scenario, the native plant resources on the site demonstrated a net present value (NPV) of $9,192 (fruit, $7,679; latex, $428; timber, $1,084). Using the second scenario, the NPV comes to $8,610 (fruit, $8,003; latex, $446; timber $161). It is important to note that in this latter scenario, fruit represents 88.2 percent, and fruit and latex together, the "minor forest products," 98.1 percent of the total present net value of the forest.

This study, comprehensive as it is, falls short for four reasons. First, the land examined is not representative of moist tropical forests in general. In addition, the study did not incorporate estimates of the values of non-marketed goods and services. The financial conclusions are circumspect in that the effect of greatly increased quantities of valued goods on market prices was not incorporated, and, finally, no allowance was made in the calculations for sustainable production over time. Despite these shortcomings, the net present value calculations for the Mishana forest demonstrate that natural forest utilization has the potential to be economically competitive with other forms of land use in the tropics. Using identical investment criteria, the NPV of the timber and pulpwood obtained from an intensively managed plantation of *Gmelina arborea* in Brazilian Amazonia is estimated at $3,184, and gross revenues from fully stocked cattle pastures in Brazil are reported to be $148/hectare/year, with a NPV of $2,960.

CASE 2. RURAL AFFORESTATION PROGRAMS IN NIGERIA

Investment projects in shelterbelts and farm forestry in arid zones in northern Nigeria demonstrate that the ecological benefits of rural

Table 7.4
COST-BENEFIT ANALYSIS OF CONSERVATION INVESTMENTS IN NORTHERN NIGERIA

Case	Yield Effect (%)	Costs Relative to Base Case (%)	Rate of Decline of Soil Fertility (%)	NPV (naira per hectare farmed)[a]	B/C	IRR (%)	Remarks
Shelterbelts							
1	20	100	1	170	2.2	14.9	Base case
2	15	110	1	110	1.7	13.1	Low yield/high cost case
3	25	100	1	221	2.6	16.2	High yield case
4	20	100	0	108	1.8	13.5	No erosion
5	20	100	2	109	1.8	13.6	More rapid erosion
6	20+	100	1	262	2.9	16.9	Soil restored to initial condition, plus yield jump
7	0	100	0	−95	0.3	4.7	Wood benefits only
Farm forestry							
1	10	100	1	129	4.5	19.1	Base case
2	5	150[b]	1	70	2.3	14.5	Low case (no "high" case assumed)
3	10	100	0	75	2.9	16.6	No erosion
4	10	100	2	60	2.5	15.5	More rapid erosion
5	10+	100	1	203	6.1	21.8	Soil restored to initial condition, plus yield jump
6	0	100	0	−14	0.6	7.4	Wood and fruit benefits only

BASED ON: Anderson 1987.

NOTE: B, benefits; C, costs; NPV, net present value; IRR, internal rate of return.

a. A 10 percent discount rate was used; 1 naira = 1 U.S. dollar.

b. This increase corresponds to a three-to-four-year lag in farmer response, plus a 10 percent cost.

afforestation programs can be translated into economic terms (Anderson 1987).

The planting of public shelterbelts and implementation of farm forestry practices can prevent soil erosion and the loss of soil fertility resulting from deforestation and loss of trees on farmland. In the more denuded areas, planting may enhance soil fertility and in areas still being cleared for agriculture, the same ecological effect can be achieved at a fraction of the cost by leaving trees standing. The overall outcome would be an increase in farm income because of the higher output of crops and livestock. The result would also be sustainable because the long-run threat to the soil's carrying capacity from erosion and from loss of nutrients (and moisture) would be reduced. In addition, there would be economically important by-products such as firewood, fuel, fruit, mulch, and fodder.

The benefits of preventing declines in soil fertility are measured by taking the present value of all agricultural outcomes from land at the present level of soil fertility and subtracting the present value of the output, assuming a decline in soil fertility. Increases in soil fertility as a result of improved moisture retention and nutrient recycling are measured by the present value of the incremental effects of afforestation on crop yield, since farm forestry and shelterbelt programs not only prevent losses in soil fertility, but may actually improve fertility. Increases in the output of livestock products (as extra dry-season fodder becomes available from the straw associated with larger crops and from trees and shrubs) are measured by the present value of the incremental livestock production. The value of the tree products such as firewood, poles, and fruit is estimated in the usual way by multiplying the amount produced by the price of the products and calculating the present value.

The net benefit of the project is the present value of the changes in net farm incomes from cropping and livestock activities plus the benefits of wood and fruit production, minus program costs. The estimated net benefits and rates of return to investment under several scenarios are shown in table A7.1 below.

The rate of return of the project when taking into consideration ecological effects provided by trees is substantially higher than the rate of return when only the benefits from wood and fruit collection are taken into account (Anderson 1987).

CHAPTER 8

Silviculture and Logging Technology for Multiple-Use Management

FOREST MANAGERS in the tropics have made enhanced regeneration of logged forests a top priority (Fox 1968). Related management objectives include the maximization of both current and future timber yields, as well as harvesting future timber yields in as short a time as possible. As noted in previous chapters, the non-timber aspects of forest management are rarely afforded adequate attention. Yet, to attain their full economic potential, forests must be managed to optimize the total package of goods and services offered, even though timber production may remain dominant. Alternative logging and silvicultural methods can have a substantial impact on the production sustainability of both timber and non-timber goods.

LOGGING METHODS AND EFFECTS

The choice and execution of logging methods can have considerable impact on the future productivity of natural forests. First, excessive mortality in seedlings may result in inadequate stocking of desirable species; second, careless or destructive logging methods may lower present timber yields; and third, the waiting period between harvests may be prolonged by reduced growth rates in the remaining stand. Each step in logging, including felling, roadbuilding, and extraction, is a potential source of damage that may seriously diminish the value of the remaining stand unless properly planned and executed. Critical components of the forest system that can be adversely affected by harvesting operations include the residual timber stand, soil stability, game animals, rare species, and watershed regulation (Marn and Jonkers 1981). With increasing labor and other costs, the logging process may become the principal or only silvicultural manipulation that a forest experiences. Logging methods must then be designed also to enhance, to the extent possible, the growing conditions for the new crop.

The first large-scale extractions of high-quality timber used manual methods for tree felling, and manpower or draft animals for the transport of logs. Relative to current practices, these methods caused only limited damage to the forest, since they minimized skid trail width, extent of understory disturbance, and soil compaction (Hamilton and King 1983, 146). Logging practices have changed drastically in the tropics since the 1930s. Now, all but the smallest timber operators extract wood mechanically, although the forestry departments of several Indian states and the Sri Lanka Timber Corporation are notable exceptions. Presently, several log extraction methods are used worldwide: (1) wheeled or crawler tractors, (2) winch-powered ground cables, (3) skyline cable systems, and (very rarely) (4) helicopters or airships.

Crawler tractors, currently used throughout the logging industry, are particularly damaging to the residual stand (Nicholson 1958, 235). Heavy crawler tractors are so powerful that little effort is required to push through a stand of small trees. Overall disturbance is therefore greatest for tractor logging, followed by winch-powered ground-cable systems (Hamilton and King 1983). Total soil disturbance from skyline logging is less than half that of ground-cable logging, due to more limited road and skid trail requirements. Aerial extraction causes the least amount of damage, but is very expensive. It is therefore cost-effective only in the removal of extremely valuable timber in areas where conventional extraction costs are quite high, or prohibited by law in order to avoid damage to valuable watersheds or other important resources.

Extensive areas of tropical forest are often unnecessarily damaged during selective harvesting operations. In a study of 45 hectares of tractor-logged forest in Sabah, Malaysia, Nicholson (1958) found that an average extraction rate of 11.5 trees per hectare resulted in logging damage to 53 percent of the remaining trees of approximately 10 cm basal diameter and over, leaving an average of 20 undamaged commercial trees per hectare. Of the damaged trees, almost half were considered to have received serious injury, i.e., they were likely to die or experience a significant decline in growth.

Damage to the forest is particularly severe when large trees with wide, spreading crowns are felled against neighboring trees, which then break and fall against additional stems. This type of damage can be accentuated by the presence of vines and woody climbers, which pull on neighboring trees when a large timber tree is felled. In the forests of Sabah, the mean stocking of climbers can exceed 1,950 per hectare (Fox 1968, 327). Experimental cutting of vines and climbers prior to logging has been shown to increase the number of undamaged and lightly damaged trees from 26 percent to 42 percent (Fox 1968, 327). Additional research in Sabah has shown that although continuous mortality of seedlings occurs in the undisturbed forest understory, the rate rises considerably following logging. Seedling survivorship in an undisturbed forest was found to be 59.5 percent over a three-year period, whereas survivorship in a recently logged stand was 13.7 percent for the same time period (Liew and Wong 1973).

To ensure sustained yields of timber in the future, stems of desirable species must survive to maturity following logging operations. A study of logging damage in Sarawak (Hutchinson 1986, 144) found that the incidence of stem injury did not differ significantly according to wood quality group. Hutchinson's findings illustrate that extraction crews often disregard the need to avoid damaging the stems of commercial species. It has been estimated that as much as one-third of logged areas typically suffer from some level of soil disturbance during harvesting operations (Ewel 1981). Additionally, the damage is rarely restricted to the soil surface, since tractors compact the soil and damage shallow tree roots, particularly when the soil is moist.

Soil compaction is especially pronounced in road building. The area occupied by roads (excluding skid trails and landings) in a normal logging operation is about 40 square meters/hectare for main roads and 400 square meters/hectare for secondary and feeder roads (Marn and Jonkers 1981, 2). Although this represents only 4 percent of the total logged area, it remains significant, since the complete removal of the vegetation and the extent of soil compaction substantially retard tree regeneration on the affected surfaces. Soil microorganisms, including the mycorrhiza that assist in nutrient uptake by trees, can be adversely affected by soil compaction and the high soil temperatures that result from a loss of vegetative cover (W. Smits, personal communication).

Intensive logging of tropical forests adversely affects animal species sensitive to changes in forest microclimates or dependent on food supplies affected by logging activities (Johns 1983). In his study of hunting in Sarawak, Caldecott (1987) found a dramatic reduction in the bearded pig (*Sus barbatus*) population that was strongly correlated with logging, although this decline was not entirely a direct result of the logging (see appendix 5.2). In addition to removing or damaging tree species on which bearded pigs depend for food, logging roads facilitated hunters' access to the forest. Given the substantial increase in shotgun ownership in Sarawak during the past few decades, this has resulted in a 65 percent decline in meat production in some forest areas. Furthermore, due to the importance of non-game animals in pollination and fruit dispersal, their loss, resulting from successive cycles of extensive logging damage, is likely to lead to a permanent reduction of biological diversity. Such an outcome could be mitigated if adjacent conserved areas are retained effectively to permit re-entry of important species.

Logging has a critical impact on watershed protection. By reducing the vegetative cover over forest soils, logging can substantially increase both surface water run-off and rates of erosion (see chapter 6). This is especially true on hill slopes, since much of the slope stability in forested areas can depend on tree roots, many of which will die and decay due to logging damage. In addition, logging debris can choke streams, and increased sediment loads can seriously damage downstream fisheries and irrigation projects (see chapters 2 and 6). Much of this damage can be prevented by keeping roads and logging activity as far from waterways as possible. Hamilton and King (1983, 150) recommend that buffer strips of at least 50 meters width be maintained in logging operations to reduce damage to streams.

CONTROL OF LOGGING DAMAGE

Measures to cope with logging damage may be either preventative or remedial. Measures to prevent damage could include silvicultural practices that precede harvesting, as well as control of logging operations to minimize damage to the remaining stand. Determining the optimal intensity of logging, improving logistical planning of log recovery, and reducing the number of roads and skid trails all constitute potentially effective steps toward prevention. In situations where extensive damage has already occurred, remedial measures such as enrichment planting must be considered.

When logging damage is not controlled, desired harvesting cycles cannot be achieved. Forest management plans often assume yields of

50 cubic meters/hectare at each felling. However, current logging practices often destroy up to 40 percent of the residual trees and kill almost 50 percent of the young growing stock (Marn and Jonkers 1981, 12). Therefore, when realistic growth rates are taken into account, it will be impossible to achieve the 25- to 35-year cutting cycle currently envisaged in many selection systems if this level of logging damage is allowed to continue.

A comprehensive study of logging methods in Sarawak's dipterocarp hill forests was conducted by Marn and Jonkers (1981). They compared the efficiency of current logging practices with that of an operation planned and executed to minimize damage and reduce overall logging costs. In the experimental block, the main trails and landings were planned on the basis of a topographic survey that identified concentrations of commercial trees. The main trails were located as close as possible to the denser stands of timber. Secondary skid trails were short, not exceeding 50 meters in length, and located wherever needed to reach logs that could not be skidded from the main trail. To make skidding as simple and efficient as possible, all trees were felled in a herringbone pattern to intersect the main trails at 30° to 45° angles. The full skidding capacity of the tractors was utilized by employing chokers and hauling two or more logs in one load.

As commonly practiced in Sarawak and throughout the tropics, logging generally involves little planning and no technical supervision. Felling begins at the landing, and proceeds into the logging block. The tractor follows behind the feller, proceeding from log to log and skidding them one at a time. Trees are cut in the direction convenient to the feller, and are thus scattered at random. Extending as they do from block to block, skid trails are usually long, steep, and winding. As a result, log skidding tends to be slow, and damage to both the logs and the remaining stand is extensive. Finally, tractor drivers frequently adopt a near-random search pattern while looking for cut logs, destroying many trees in the remaining stand in the process (Marn and Jonkers, 1981, 5).

An analysis compared the efficiencies of the two logging methods used in the Sarawak experiment. The cost of felling remained the same, indicating that directional felling was achieved without any increase in cost. Planning the main skid trails, and establishing them prior to felling, increased logging costs slightly, while skidding costs were reduced by approximately 23 percent (Marn and Jonkers 1981, 6). In terms of machine efficiency, experimental logging outperformed the current system by 36 percent (i.e., 20.0 cubic meters/hectare skidded against 14.7 cubic meters/hectare over an average distance of 290 meters) (Marn and Jonkers 1981, 6). Damage done in the experimental block was substantially lower than that resulting from current methods. Only 17.1 percent of the experimental block was classified as temporarily opened space, compared with 30.4 percent in the control block, even though the area occupied by skid trails was the same. Also, uprooted and broken trees were twice as common in the control plot. Assuming that 35 percent of the damaged trees will not recover, the total number of commercial trees lost due to logging was approximately 40 trees/hectare in the experimental block, compared with 60 trees/hectare in the control plot (Marn and Jonkers 1981, 8). Hence, the experiment demonstrated that adopting a few simple control measures can both reduce logging damage substantially and reduce overall extraction costs in the process.

ENRICHMENT PLANTING

In forests where the stocking of seedlings and saplings of desirable species is inadequate for natural regeneration, either because of inherently low seedling survival or as a result of destructive logging methods, the available remedial measures are limited. One commonly invoked option is to do nothing—to wait in the hope that subsequent fruiting of the desired species, aided by the recent opening of the forest canopy by logging, will result in a healthier seedling population. This option has a chance of succeeding in situations where parent "seed trees" are left in sufficient density to ensure adequate coverage of the opened forest by seedfall, and if climber and herb competitors are not so dense as to preclude seedling establishment. Natural regeneration of this sort was found to succeed in some cases in Nigerian forests, especially when the understory was cleared of competitors prior to or shortly after logging (Kio and Ekwebelam 1987).

In many cases, however, natural regeneration cannot replace the seedlings and saplings lost through logging damage. In some regions, including most of Asia and much of Africa, natural regeneration of commercial species depends largely on surviving, established seedlings, since fruiting is irregular and therefore unlikely to occur before a buildup of weed competitors following logging (Whitmore 1984; Kio 1987; see chapter 3). Consequently, the destruction of the preexisting seedling and sapling layer, combined with extensive damage to pole-size trees, effectively removes these species from the next cycle of forest growth. In these cases, artificial regeneration through enrichment planting is perhaps the only practical means of restoring the desired species to the forest stand.

The success of enrichment planting varies enormously with the methods used, the species planted, and the extent and quality of post-planting care. The practice has been applied to a variety of forest types in both Asia and Africa, with an emphasis on the regeneration of dipterocarp and mahogany species, respectively (Whitmore 1984; Nwoboshi 1987). An important exception to this occurs in limited areas of the Philippines and Malaysia, where fast-growing species are planted in exposed areas along logging roads, skid trails, and landings in an attempt to reforest these heavily degraded sites (Appanah and Salleh 1987). In Indonesia, attempts are being made to plant both fast-growing pioneer and slow-growing mature-phase species on the same site (Spears 1987). This technique is likely to transform the forest composition and structure substantially, and is therefore treated as a plantation method in this study (see chapter 9).

Two methods of enrichment planting predominate: (1) line planting, in which seedlings are planted out in corridors, generally cleared of much of the overhead; and (2) group planting, in which groups of seedlings are planted in a naturally occurring or artificially created canopy gap (Kio 1987). Seedlings are occasionally planted beneath closed or partly open canopy, although this is less common, since a principal component of the method is to place seedlings in favorable light environments. Post-planting care consists primarily of climber cutting and weed control, although in practice this care is often neglected (Schmidt 1987).

In general, the success of enrichment planting in promoting the regeneration of commercial tree species has been limited at best. Seedlings often fail to establish, and those that do frequently exhibit poor growth or are overwhelmed by climbers and weeds (Liew and Wong 1973; Kio and Ekwebelam 1987). Consequently, the efficacy of enrichment planting in natural forests has been widely questioned, with some managers concluding that it is unlikely to be cost-effective in many situations due to high labor requirements and/or poor seedling survival and growth (OTA 1984; Asabere 1987).

However, a review of previous attempts at enrichment planting suggests that most failures are due to improperly applied methods, rather than a flaw in the concept itself (Kio 1987). Specifically, enrichment planting often encounters the following difficulties:

1. pot-bound seedlings with low root/shoot ratios are frequently planted, exposing them to considerable moisture stress;
2. the canopy above the planting site is often excessively reduced, creating hot, dry conditions hostile to the mature-phase species planted;

3. supervision of seedling planting operations is inadequate, result-ing in high immediate mortality; and
4. follow-up maintenance of the site is neglected, so that climbers and weeds are able to out-compete the seedlings for light, water, and nutrients.

Given sufficient attention to these problems, however, it seems likely that enrichment planting could provide an acceptable means of en-hancing the regeneration of desired species in inadequately stocked forests. Enrichment planting has generated high success rates, achieved in both moist deciduous and evergreen forests, in Karnataka, some other Indian states, and Sri Lanka over many years (M. Ashton, personal communication, 1987).

A particularly promising technique involves the enrichment planting of forest patches disproportionately damaged by logging (i.e., road-sides, skid trails, and landings). Even in cases where harvesting opera-tions are carefully controlled, the creation of substantially denuded sites is unavoidable. In the absence of enrichment planting, such areas will be lost to commercial wood production in the next growing cycle. Enrichment planting on these sites could partially restore their produc-tivity. It is essential, however, that the species used be adaptable to the harsh conditions found on degraded sites. This will generally mean pioneer or building-phase species, whose seedlings are often adapted for growth in forest gaps or on disturbed soils. As described in chapter 9, such sites can subsequently be underplanted with more valuable, mature-phase species once the pioneer or building-phase trees are established and growing.

Non-Timber Products

As described in chapter 3, the silvicultural treatment of tropical forests is generally directed toward guilds of species, such as the pioneer or mature-phase groups, rather than toward individual species. This in-cludes both timber and non-timber trees, as well as most vines and understory plants. Therefore, proper management of the forest for timber production, such as control of logging activities, will generally benefit the non-timber species of the same guilds as well. Exceptions to this rule will occur when the following conditions prevail: (1) the non-timber species belong to different guilds than the timber species (e.g., understory herbs of the deep forest); or (2) silvicultural practices, such as extensive improvement felling, result in disproportionate mortality to non-timber plants. Of the two, the latter is the more serious source of conflict, since most of the important non-timber species (i.e., rattan, bamboo, and many fruit-producing vines and shrubs) belong to guilds that benefit from the limited canopy removal associated with selective logging.

The impact of silvicultural practices on non-timber species depends on the nature of the practice and the extent to which non-timber goods and forest services are emphasized in its implementation. As indicated in chapter 3, the use of extensive poison-girdling in uniform shelter-wood systems can result in the elimination of potentially valuable species, both timber and non-timber. Liberation thinning, at least as it is practiced in Sarawak, substantially reduces this threat, since it calls for poison-girdling primarily those trees in direct competition with selected timber trees. In addition, the recognition of the importance of fruit trees to the local population in Sarawak has resulted in an injunction against poison-girdling important trees such as durians, mangosteens, and mangoes (S. Tan, personal communication, 1987).

The animal populations of tropical forests, including many of the game species, are often dependent on a limited number of tree and vine species for fruit and seeds. Of particular importance are plant species that produce fruit year-round, providing an important source of food between periods of abundant fruiting (McClure 1966). Some of these plants, such as figs in Southeast Asia, are essential for maintaining some animal populations during times of low fruit availability in the remainder of the forest (Leighton and Leighton 1983). In addition, fruit produced by vines of the *Annonaceae* constitute an important source of food for many primates during times of scarcity (L. Curran, personal communication, 1988). It is important that these "keystone" species be protected from logging damage or poison-girdling if animal populations are to be preserved at reasonable levels.

CONCLUSION

In order to realize the full economic potential of tropical forests, the evidence available indicates that they must often be managed for forest services and non-timber goods as well as timber. From a technical standpoint, a major threat to future supplies of both timber and non-timber goods from protected tropical forests is unmanaged logging activity.

Historically, selective logging based on hand-sawing and extraction by draft animals had relatively little impact on forest integrity. The advent of chainsaws and crawler tractors has substantially changed this pattern. Careless tree felling, excessive use of roads and skid trails, and disregard for damage to understory vegetation often leads to high rates of mortality in pole-size trees, saplings, and seedlings. The result is poor regeneration of desired species, implying a severe reduction in future timber potential. In addition, damage to non-timber species results in substantial reductions of non-timber goods, including important plant foods and game. Yet studies show that logging damage can be substantially reduced through logistical planning ahead of time, careful placement of logging roads and skid trails, and directional felling to facilitate log extraction, while simultaneously reducing harvesting costs.

Enrichment planting may be the only reliable means of regenerating seriously degraded natural forests within a reasonable period of time. To date, however, the success of enrichment planting has been disappointing, with failures attributable to the mismatching of species and understory environment, inadequate supervision of planting, and insufficient follow-up maintenance. Properly implemented, however, enrichment planting holds promise for restoring degraded forests to productive status. Particularly intriguing is the possibility of enrichment planting on denuded patches within logged forest.

Generally, proper methods of forest management for timber production will benefit non-timber species as well. Exceptions occur when non-timber species belong to plant guilds that are adversely affected by canopy thinning, or when silvicultural practices are employed that target valuable non-timber species for removal. In some countries, such as Malaysia, important non-timber species are exempted from practices such as canopy thinning.

CHAPTER 9

Plantation Forestry

THE DIFFICULTIES associated with the management of natural tropical forest have led some forest managers to conclude that silvicultural methods of enhancing natural regeneration following logging are often ineffectual, and therefore are rarely cost-effective (Leslie 1977). This is particularly true if the better forest sites, such as those occupying good soils in lowland areas, are ultimately destined for conversion to agriculture or tree-crop plantations. In view of the diminishing extent of accessible land available for forestry, rates of logging well beyond maximum sustainable yield in virtually all regions of the tropics, and the variable response of natural forests to silvicultural treatment, forest plantations are widely regarded as an economically attractive alternative form of forestry investment (Spears and Ayensu 1985; WRI 1985).

TYPES OF PLANTATIONS

The approximately 12 million hectares of forest plantations established in the tropics (Lanly 1982) can be classified into three major categories: (1) short-rotation industrial plantations; (2) short-rotation nonindustrial plantations; and (3) longer-rotation industrial plantations (see tables 9.1 and 9.2).

The first and largest category consists of about 5 million hectares of short-rotation industrial plantations, producing paper pulp, wood chips, and low-density sawlogs. The species chosen for these plantations are selected from a relatively short list of fast-growing trees of the pioneer and building-phase guilds, including species of *Acacia, Paraserianthes (Albizia) falcataria, Anthocephalus, Araucaria, Eucalyptus, Gmelina, Leucaena,* and *Pinus,* among others (Evans 1982; Whitmore 1984; see chapter 3 for guild definitions). Roughly one half of these

171

Table 9.1
TRENDS IN PLANTING RATES OF PLANTATIONS
IN THE HUMID TROPICS
(IN THOUSANDS OF HECTARES)

	1931–40	1941–50	1951–60	1961–65	1966–70	1971–75	1976–80
Africa	1.14	11.38	48.57	43.94	69.96	95.96	123.42
Asia/Pacific	6.10	4.10	8.30	373.30	380.43	852.60	1,549.84
Latin America	12.15	20.80	401.80	102.55	423.20	1,406.85	1,935.65
TOTAL	19.39	36.28	458.67	519.79	873.59	2,355.41	3,608.91

BASED ON: Grainger 1986.

172

Table 9.2
AREAS OF DIFFERENT TYPES OF FOREST PLANTATIONS
IN THE HUMID TROPICS, 1980
(IN THOUSANDS OF HECTARES)

				Industrial Hardwood	
	All	*Industrial*	*Hardwood*	*Sawlogs*	*Gmeline arborea*
Africa	688.9	407.3	282.8	200.7	66.8
Asia/Pacific	2,618.0	1,705.5	1,174.5	1,086.1	0.0
Latin America	4,430.3	2,467.5	947.1	92.8	71.0
Humid Tropics	7,737.2	4,580.3	2,404.4	1,379.6	137.8

BASED ON: Grainger 1988.

plantations are in Latin America, specifically in Brazil, with softwood species predominant. Asia holds 35 percent of these plantations, and Africa accounts for only 15 percent of the total (Lanly 1982). The area under short-rotation industrial plantations is expanding rapidly, with an estimated 0.5 million additional hectares planted annually in all developing countries, excluding China (Spears and Ayensu 1985).

The second category includes about 4.5 million hectares of non-industrial plantations, composed primarily of *Pinus* and *Eucalyptus* species grown for charcoal and fuelwood (Lanly 1982). The majority of these plantations are located in Latin America (46 percent) and Asia (36 percent), where they produce mostly industrial charcoal and household firewood, respectively. In Africa, most plantations of this type are found in the drier tropical regions, where the scarcity of firewood is particularly acute (Lanly 1982). Many of the fuelwood plantations in Africa and Asia are established in conjunction with community development projects (WRI 1985). Relatively more effort has gone into expanding these types of plantations than any other category in recent years (Lanly 1982).

The third and smallest category consists of approximately 2.5 million hectares of longer-rotation plantations of climax species, producing high-quality sawn timber and veneer. Of these plantations, which represent only 20 percent of all forest plantations in the tropics, the large majority consists of older teak plantations in India and Indonesia. In Africa, the roughly 300,000 hectares of timber plantations are nearly evenly divided between teak and indigenous mahogany species (Lanly 1982). Few timber plantations have been established in tropical Latin America, where there is a strong emphasis on industrial plantations of fast-growing species (McGaughey and Gregersen 1983).

EXISTING PLANTATION DESIGNS

A review of the distribution and nature of tropical forest plantations established within the past few decades reveals five salient characteristics critical to the ability of plantations to compensate for the loss of natural forests. These characteristics are:

Emphasis on Fast-Growing Species

Perhaps the clearest recent trend in tropical plantation forestry is a strong shift toward industrial plantations based on fast-growing softwood and light hardwood species (Lanly 1982). This trend is most pronounced in Latin America, where large-scale industrial plantations, such as the Jari Project in Brazil, have been established under the stimulus of substantial tax incentives (McGaughey and Gregersen 1983). Even in Asia and Africa, with their longer histories of experimental plantation development, plantations of fast-growing industrial trees are being established at a much faster rate than plantations of slower-growing timber trees (Lanly 1982). Recent calls for increased investment in tropical forest development stress the potential role of industrial plantations of fast-growing species in tropical wood production (Spears and Ayensu 1985).

The growing emphasis on fast-growing species reflects a desire on the part of forest managers for a plantation that:

1. has a short rotation, providing a rapid return on investment;
2. has a simple stand structure, facilitating silvicultural treatment;
3. provides a uniform product; and
4. can be harvested in a single felling (Evans 1982; Hartshorn 1983).

The first two attributes are particularly important in community forest plantations, since the need for fuel and construction materials is generally already acute by the time plantations are considered, and management skills are often limited (Evans 1982; WRI 1985).

There can be little doubt that plantations of fast-growing species successfully provide a variety of local and industrial wood products in many tropical countries (Spears and Ayensu 1985; WRI 1985). However, a large percentage of short-rotation plantations yield primarily non-timber products such as fuelwood, paper pulp, and wood chips (Lanly 1982). In situations where fast-growing trees are grown for plywood or light construction timber, the comparatively low quality of these products is likely to preclude their substitution for the higher-density timber derived from natural forests, at least in the international market. Short-rotation plantations cannot, therefore, be equated with plantations of slower-growing trees with respect to high-quality timber production. Nor can they directly compensate for the loss of productive natural forests.

Monoculture Plantations and Mortality Risks

A second principal feature of virtually all tropical forest plantations is that they are grown as monocultures, or large blocks of a single species, similar to the manner in which tree-crops such as rubber and oil palm are grown (Lanly 1982). A major advantage of monoculture design is its greatly simplified stand structure, making both silvicultural treatment and harvesting operations easier (NRC 1982). In addition, monoculture plantations offer the possibility of a predictable and uniform supply of wood, in contrast to the more variable supply of diverse wood types obtained from natural forests (Evans 1982). This feature of monoculture plantations can be particularly important to locally-based wood industries, which depend on the consistency of regional production (Spears 1979).

A number of potential problems associated with monoculture plantations, however, cast doubt on their long-term sustainability in the humid tropics. Perhaps the most important of these is the absence of species diversity, in contrast with the extremely high diversity generally found in tropical forests. Although the absence of diversity in plantations may be a desirable marketing feature, from a biological standpoint it introduces a serious risk of inordinately high tree mortality from insect pests and diseases.

Several ecological studies indicate that the high species diversity of tropical moist forests is maintained in part by mortality agents, particularly host-specific insects and diseases, which differentially attack plant species occurring in high densities (Augspurger and Kelly 1984; Clark and Clark 1984; Hubbell and Foster 1987). The absence of a dormant period (cold or prolonged dry season) in much of the humid tropics allows the uninterrupted reproduction of insects or disease organisms in plant populations in which individuals occur in sufficient proximity to allow the transmission of the problem from one plant to another. According to one set of hypotheses, these mortality factors act to inhibit the regeneration of the more common tree species as their numbers approach some critical density, thereby preventing them from excluding other, perhaps less competitive species from the forest community (Janzen 1970; Hubbell 1980). Although most of the evidence for these hypotheses is focused on the seedling stage of tree populations, at least one large-scale demographic study in Panama suggests that density-dependent mortality effects extend to the juvenile and adult stages as well (Hubbell and Foster 1987).

Evidence of the intense pressure placed on tropical trees by herbivores can be found in the high concentrations of defensive secondary chemical compounds in many tropical plants, of which the bulk of the world's spices are an example (NRC 1982). Additional direct evidence is provided by the frequent outbreaks of insect pests and diseases in tropical crop plantations. For example, recent attempts to introduce large-scale monoculture plantations of indigenous fruit trees in West Malaysia have been frustrated by insect attacks, while in nearby villages traditional tree gardens containing the same species planted in mixture have thrived for centuries (P. Ashton, personal observation, 1987). Nor are such outbreaks of insect attacks confined to artificial monocultures alone. In Sarawak, large gaps are occasionally created in natural peat swamp monocultures of *Shorea albida* through defoliation, apparently by tussock moth caterpillars. In contrast, the formation of large gaps due to insect-caused mortality in unrelated trees is virtually unknown in species-rich, undisturbed tropical forests (Whitmore 1984).

The success of eucalyptus, pine, and teak monocultures in the subtropics and seasonally dry tropics is often taken as evidence for the viability of forest plantations in the tropics in general. However, as noted above, the success of plantations in these areas is not necessarily transferable to the perennially moist climate of the humid tropics, where the lack of a dormant season can facilitate the buildup of pests and diseases in monospecific stands (UNESCO 1978). Examples of the potentially severe constraint this poses for plantation forestry in the aseasonal tropics can be found in the difficulties presented by leaf-cutter ants in the Jari Project in Brazil (G. Hartshorn, personal communication, 1987), the occurrence of fungal heartrot in *Acacia mangium* plantations in Sabah, Malaysia (A. Moad, personal observation, 1987), and the current lethal spread of the neotropical bug *Heteropsylla cubana* through *Lencaena polycephala* plantations in the western Pacific.

The existence of large-scale plantations of tree-crops in the moist tropics (rubber, cacao, coffee) demonstrates that these problems can be overcome, but often only through the frequent application of chemical sprays to control insects and diseases. Although such practices would probably be equally effective in industrial wood plantations, it is likely in the case of forestry plantations that the cost of spraying would be prohibitively expensive in comparison with the value of the crop being protected. An alternative, more cost-effective form of control might be obtained by reducing the density of the host species. An excellent example of this phenomenon can be found in Brazil, where rubber extraction from trees occurring naturally in low density in undisturbed forest continues to provide employment for thousands of rural people (Allegretti and Schwartzman 1986), whereas attempts at monoculture plantations were eventually abandoned due to losses from leaf blight (NRC 1982).

Another way of avoiding host-specific insects or diseases in plantations, at least in the short-to-medium term, is to grow trees outside their natural range in an environment where such pests are unlikely to have evolved. The tremendous success of rubber plantations in Southeast Asia is partly attributable to this type of release from native pests (NRC 1982). An example involving a timber-producing species is found in Sri Lanka, where the experimentally planted mahogany *Swietenia* appears to be thriving (P. M. S. Ashton, personal communication, 1987), whereas plantations of this tree are plagued by shoot borers in its native Latin America (NRC 1982). The avoidance of co-evolved pests does not, however, preclude the possibility of local generalist pests being capable of attacking the introduced tree. Nor does it avoid the long-term risk of the accidental introduction of co-evolved pests, as has frequently occurred with tropical tree-crops such as cacao and coffee.

Spatial Arrangement in Plantations

An additional constraint on production in monoculture plantations involves tree architecture and the optimal use of space. Since all trees in an even-aged monoculture will be of roughly the same size, and will share identical crown architectures, each tree will be in direct competition with its neighbors for canopy space. The potential for packing trees into a plantation will therefore be determined primarily by only two dimensions of the plantation—length and width. The varying sizes and architectures of the numerous species found in natural forests extends the space available for tree crowns into the added dimension of height as well. This permits a greater photosynthetic leaf area to occupy a given plot of land, since the added dimension of height allows extensive overlap of tree crowns. The result is a greater density of tree stems than would occur if the tree crowns were confined to a single layer. An intuitive feeling for this effect can be had by comparing an overview of a monoculture plantation, which gives the appearance of a relatively uniform carpet of tree crowns, and a natural forest, which is both uneven and has greater overall crown depth.

There is a limit to the tree-packing benefit afforded by overlapping tree crowns, since the shading of one tree's crown by another will affect the growth of the shaded tree. Also, since most trees of the wet tropics have relatively shallow roots (Whitmore 1984), there is limited opportunity for root overlap, and therefore an increased probability of root competition as trees are more densely packed. Nevertheless, the inclusion of trees with more than one crown architecture and growth rate holds potential for increasing wood production on a given area of land.

In addition to expanding the volume of the forest canopy, the inclusion of several species in a plantation introduces the possibility that the survival or growth of one species may be facilitated by the presence of another. For example, many timber trees of the mature-phase/light hardwood guilds, including many dipterocarps of Asia and most of the mahogany species of Africa and Latin America, will not establish or grow well as seedlings in full sun (Whitmore 1984; P. Kio and G. Hartshorn, personal communication, 1987; P. M. S. Ashton and P. S. Ashton, 1992; P. M. S. Ashton, in press). Instead, they generally establish in small gaps or beneath closed canopy, growing to maturity when gaps form to allow additional sunlight to reach the forest floor. In the successional sequence that occurs in the natural regeneration of a large gap or cleared site, the presence of pioneer or building-phase trees is often necessary to ameliorate the harsh environment of a large, open space before mature-phase species can become established. The potential for designing tree plantations that incorporate species of more than one successional guild (i.e., mature-phase/light hardwoods beneath building-phase trees) to emulate this process of successional facilitation remains largely unexplored.

Another example of the facilitation of one species by another concerns the beneficial effects that nitrogen-fixing and deep-rooted trees have on soil nutrient cycling. Through symbiotic associations with root-nodule bacteria many tropical trees, including several important pioneer and building-phase species in the Leguminoseae and Casuarinaceae, are capable of directly fixing atmospheric nitrogen (Whitmore 1984). In addition, species with especially deep roots are able to capture nutrients from areas deep within the soil profile that are not available to most tropical trees. The potential contribution of trees with these root characteristics to overall forest growth is twofold. First, by utilizing reservoirs unavailable to many species, competition for nutrients is reduced. Second, these trees may act as nutrient "pumps" by bringing in nutrients from outside the system (atmosphere or subsoil) and releasing them through their leaf litter (Raintree 1986).

Harvesting Strategies

In virtually all cases, industrial plantations of fast-growing species are managed as monocyclic shelterwood systems in which large blocks of forest are clear-cut on a short (10 to 20 year) rotation and then re-planted with another cycle of trees. There are two problems associated with this harvesting strategy in the humid tropics. The first derives from the short harvesting rotations of fast-growing tree plantations, while the second involves the high levels of soil degradation inherent in clear-cutting operations.

The soils of the moist tropics often exhibit very poor nutrient-holding capacity, with the above-ground biomass holding a large proportion of a forest system's nutrients (Whitmore 1984). Consequently, when a large volume of the standing biomass is removed, a significant proportion of the system's nutrient reservoir is removed with it; phosphorous, in particular, tends to be lost along with the removal of biomass (NRC 1982). This is especially true of areas allocated for plantation forestry, since most sites with better soils have already been or are likely to be converted to agriculture or tree commodity crops. It is therefore probable that applications of chemical fertilizers will be required to grow plantation trees on the same soils beyond the first few harvests (NRC 1982). If the harvest rotation is short, as expected in plantations of fast-growing trees, sooner and more frequent applications of fertilizer are likely to be needed. This holds potentially serious implications for both the biological and economic sustainability of short-rotation industrial and fuelwood plantations.

Given the use of clear-cutting as the principal means of harvesting in monoculture plantations, a higher level of soil degradation than would occur under selective harvesting seems inevitable. Large expanses of soil will be exposed periodically to direct rainfall and sunlight, resulting in substantially increased erosion and higher soil temperatures capable of killing the root mycorrhizal symbionts that aid in nutrient uptake by trees (Whitmore 1984; Spears 1987; W. Smits, personal communication). In addition, the often thin layers of organic litter and humus will be liable to increased rates of decomposition, reducing soil nutrient-holding capability. The more frequent the harvesting, the more pronounced these effects will be. If one assumes that most forest plantations are likely to be established on poorer soils, the potential for plantations of fast-growing monocultures to contribute to the degradation of soil structure is therefore substantial.

Conversion of Natural Forests

A final, critical feature of plantation forestry in the tropics is the widespread tendency to establish industrial plantations on lands excised from existing tracts of natural forest. With the exception of watershed restoration projects or fuelwood plantations (Spears 1982), deforested sites are rarely selected for plantation forestry. In some cases, plantations are established on secondary forest sites that exhibit poor regeneration of desired species. More commonly, however, plantations are located on areas of natural forest that possess high, marketable timber volume and good regeneration characteristics.

In some cases, the avoidance of poorly forested areas reflects the recognition that plantations are less likely to succeed on degraded sites than in areas converted from existing forest, where soil conditions are more favorable. A more compelling reason for the conversion of natural forests may lie in the opportunity for the plantation owner, whether private or public, to reap the often substantial profits involved in harvesting a valuable block of forest. Throughout the tropics there are forest plantations that were ostensibly created to provide wood products, but whose principal effect has been the reduction of natural forests, often at substantial profit to concession holders. The frequent absence of adequate testing of plantation designs prior to their widespread application can also be attributed, at least in part, to the profitability of forest conversion. In plantations where the majority of the profit is derived from initial forest conversion, rather than the subsequent production of wood, there is little financial incentive to insure that the plantation will in fact succeed.

In some cases, therefore, the creation of industrial forest plantations is used more as a means of legally excising blocks of protected forest reserves, and thereby accelerating logging rates, than as an alternative method of producing wood products. In situations where productive, natural forest is converted to industrial plantations this conversion generally represents a net loss of timber-producing capacity, since fast-growing plantation species generally yield different products. Of the plantations recently established or projected for the near future, an increasing proportion consists of the short-rotation, non-timber-producing categories (Lanly 1982).

The loss of non-timber products resulting from the conversion of natural forests to industrial plantations is rarely addressed in economic assessments of plantation designs. In theory, the loss of such benefits as fuel, game, and other non-timber goods resulting from the conversion of natural forests should be included as an opportunity cost in project cost/benefit analysis. In practice, these losses are often given little or no consideration in forest development plans, in part because they are not incurred by the concessionaire or the government forestry agency, and also because these goods often contribute little to export earnings. Yet, as indicated in chapters 5 and 7, these products can be quite important in efforts to maximize the total economic return from forest lands, while increasing the social equity of its distribution and thus the security of forest investments.

The conversion of natural forests to forest plantations will inevitably involve an initial loss of forest services due to the initial clearing of the vegetative cover. The subsequent provision of services will depend in large measure on the care with which the plantations are managed. It may approach that of the original forest, and will certainly be higher than alternative land uses such as agriculture or pasture (Spears 1979; NRC 1982). Consequently, with the exception of genetic conservation, loss of services resulting from converting natural forests to plantations may be relatively minor. In the case of plantations established on deforested land, however, the forest services obtained by reforestation constitute a net increase in benefits, since these services are created on land where they did not previously exist. Depending on site conditions, the species selected, and the nature of downstream land use, the benefit from these added services could be substantial, and might well outweigh any additional costs associated with establishing a plantation on a degraded site.

A careful accounting of the costs and benefits associated with forest plantation establishment can make a major difference in optimal site selection. The loss of timber and non-timber products associated with forest conversion, and the potential for creating additional services by locating plantations on degraded sites, are strong reasons for using deforested land rather than converting natural forest for the establishment of plantations.

MIXED-SPECIES PLANTATION DESIGN

The difficulties associated with industrial plantations described above cast considerable doubt on the long-term viability of short-rotation, monoculture forest plantations in the humid tropics. In addition to increased risk from insects and diseases, the possibility of soil nutrient depletion and structural degradation, and the inefficient use of available canopy space, the majority of forest plantations are simply not growing wood of sufficient quality to compensate for the loss of natural forests.

Mixed-species plantations present a possible means of overcoming many of the biological difficulties encountered by monocultures, while also producing the high-quality timber currently obtained almost exclusively from natural forests. The exact nature of these plantations would depend on the biological, social, and economic conditions of the region in which they are established, but would share a number of the following characteristics:

1. high genotypic variation, through the inclusion of a number of tree species and populations, to reduce the risk of insect pest and disease outbreaks;

2. a design that combines trees of different growth rates, heights, crown architecture, and rooting depths to optimize the use of available space;

3. the use of fast-growing, marketable trees (*Paraserianthes* [*Albizia*] *falcataria*, *Gmelina*, *Artocarpus*, *Terminalia*) as a cover crop for slower-growing, high-quality timber species (mahoganies, dipterocarps) that benefit from shade as seedlings;

4. a polycyclic harvesting strategy that allows the removal of some trees while leaving a residual forest cover to protect the soil from excessive erosion or depletion of organic material;

5. establishment on degraded forest or abandoned land to avoid the opportunity costs associated with the conversion of natural forests, and to increase the net benefit of plantations through the creation of additional forest services such as soil protection and watershed control;

6. the inclusion of nitrogen-fixing trees and trees with deep root systems to aid in the buildup and recovery of soil nutrients, particularly on degraded sites; and

7. the inclusion of non-timber products, such as cane, latex, fruit, etc., to increase the total economic return from plantations by providing a continual flow of goods, and to broaden the distribution of goods and services to include local communities, thereby enlisting their support for plantation protection.

MIXED-SPECIES PLANTATION MODELS

Few examples of commercial mixed-species forest plantations exist in the humid tropics. The majority of those that do exist were established several decades ago, and the records of their design and performance are often either missing or incomplete. Several experimental trials of mixed-species plantations, however, indicate the potential advantages and sustainability of this type of plantation design.

In his forest regeneration trials in Nigeria during the 1930s, J. D. Kennedy established approximately 50 hectares of experimental plots in which 11 African mahogany and light hardwood species were interplanted with agricultural crops, and preserved following the abandonment of cultivation. In other sites, seedlings were planted in both pure and mixed stands on partially cleared forest plots (Lowe and Ugbechie 1975). There appears to have been little subsequent maintenance of these plots beyond protection from agricultural encroachment, with the result that a number of seedlings failed to establish (Kio 1987). However, re-measurement of the surviving trees in 1970 showed high growth rates in many species, particularly the light hardwoods. For example, *Terminalia superba* exhibited average annual diameter increments of over 2.0 cm/year, which is more than twice its average growth rate in natural forest (Lowe and Ugbechie 1975). Growth rates of some of the mahogany species, *Khaya ivorensis* in particular, were also higher than normally encountered in the natural forest (Kio 1987).

Throughout Nigeria, approximately 43,000 hectares of mostly native-species timber plantations have been planted (Kio 1987). A common problem encountered in African mahogany monoculture plantations is high rates of insect and disease attack, a difficulty that does not appear to have severely affected Kennedy's mixed-species plots (P. Kio, personal communication, 1987). Given the difficulty of encouraging the natural regeneration of most mahogany species in logged forest, the use of mixed-species plantations may represent an attractive method of insuring the future production of mahogany timber in many parts of Africa (Asabere 1987; Kio 1987).

From 1927 to 1940, 10 plots of a few hectares each were established as dipterocarp plantation trials near Semengoh Forest Reserve in Sarawak, Malaysia. Maintenance of the plots consisted of weed and climber cutting, selective thinning of competing saplings, and limited felling or girdling of overtopping trees. Analysis of the plot records shows average growth rates that are three times the average observed in primary forest and twice the average in secondary forest for the six species tested (Tan et al. 1987). Unfortunately, the plots were established as monocultures, precluding the testing of any possible benefits derived from mixed-species planting. However, it is likely that the plots are too small to register any negative effects of monoculture planting in any case. The principal result of the trials is to demonstrate the substantial increase in growth rates in native species possible under plantation conditions. Similar results have been observed in experimental dipterocarp plantations at Kepong, peninsular Malaysia (S. Appanah, personal communication, 1987).

Further trials of dipterocarp plantations are currently being initiated in both Sarawak and peninsular Malaysia, with plans to include stands of dipterocarps grown in mixture with fast-growing trees (E. Chai, personal communication, 1987; J. Racz, personal communication, 1987). Large-scale plantation trials of dipterocarps and other mature-phase species established beneath a cover of building-phase trees have been suggested as a part of a major reforestation program in Sabah (Moad 1986). Through experiments such as these, controlled tests of a variety of mixed-species plantation designs, including an evaluation of their principal silvicultural and economic characteristics, would be possible.

W. Smits has proposed a model for enrichment planting of dipterocarps in logged forest in East Kalimantan, Indonesia, which approaches plantation forestry in its intensity of management and modification of forest composition and structure. In this model, heavily logged dipterocarp forest is planted twice in a period of a few years, initially with an overstory of building-phase species, then with dipterocarps propagated from cuttings and inoculated with root mycorrhiza, using a technique developed by Smits. In theory, the cover of fast-growing trees can be harvested for timber or paper pulp after 15 years, generating a short-term, partial return on investment. The dipterocarps, released from both shade suppression and root competition by the removal of the cover crop, are then allowed to grow to harvestable size, which would be reached in about 50 years. Limited trial plots of this system have already been established, with additional trials anticipated in the near future (Spears 1987).

Economic analyses of the Smits model, necessarily based on hypothetical planting costs, yields, and prices, have been conducted by Spears (1987) and Sedjo (1987). Using a discount rate of 10 percent, Spears projects a positive net present value (NPV) for the Smits model which exceeds that of natural forest management, and approaches that of industrial plantations. However, the major portion (74 percent) of this NPV is derived from the initial logging of the natural forest, with the remainder coming from the harvest of the cover trees (20 percent) and the dipterocarps (6 percent). If the value of the initial timber harvest is deleted from the model, as would be the case for plantations established on degraded sites, the amount and relative proportions of the NPV would be substantially altered (J. Spears, personal communication, 1988).

Sedjo's analysis of the Smits model does not include the income derived from the initial logging of the forest, uses higher costs for enrichment planting, and assumes that the sale of the fast-growing trees will equal the cost of their removal. Based on these less favorable assumptions, Sedjo projects a net present value of (−)$347/hectare at a 10 percent discount rate and (+)$196/hectare at a 6 percent discount rate, substantially less than the potential NPV obtained by Spears. In comparison with his best-case scenarios of natural forest management and industrial plantations, Sedjo finds that the Smits model yields much lower returns. However, as pointed out in this study, assumptions concerning the performance and sustainability of the forest and plantation systems currently in use are severely constrained by a lack of reliable data (R. Sedjo, personal communication, 1988).

As is the case for any plantation design that includes more than one successional guild of trees, a major unanswered question in the Smits model concerns the technical feasibility of harvesting a layer of pioneer or building-phase trees without causing excessive damage to the remaining mature-phase trees. Additional information concerning the costs of enrichment planting and site maintenance, expected growth rates, and probable rotation periods are needed before the model can be properly evaluated in economic terms. However, both Spears and Sedjo find sufficient evidence of the potential productivity of the Smits model to recommend additional experimentation (J. Spears and R. Sedjo, personal communication, 1988).

Non-Timber Goods and Plantations

The ability of non-timber goods to significantly enhance the net benefit derived from forest plantations depends on a variety of factors. These include the silvicultural requirements of the species involved, the cost of their establishment and maintenance, the yield and value of the goods produced, and the socioeconomic characteristics of the region in which the plantation is located.

A critical social factor influencing plantation design is local population density, since this will in large part determine: (1) the availability and cost of labor for plantation operations, (2) the need for and ability to utilize non-timber forest goods, and (3) the intensity of use, and hence the opportunity cost, of land. In areas where population density is low, the exclusive production of high-quality timber on long-rotation plantations may be tenable. In areas of higher population density, however, there will be pressure for more intensive forms of land use which provide both greater and more frequent returns, in the form of goods and/or employment opportunities, than might reasonably be expected from timber plantations. Within the context of plantation forestry, one solution to this dilemma is to cultivate fast-growing trees in industrial plantations in order to shorten the payback period. The potential difficulties involved in applying this solution in the humid tropics have already been outlined in this chapter. An alternative solution is to develop forest plantations in which the longer-term production of high-quality timber is balanced with the short-term production of industrial wood and non-timber goods to optimize the use of available labor, land, and capital on a sustainable basis.

In areas where labor is relatively inexpensive, the production of non-timber goods could add substantially to the value of forest plantations by providing a steady flow of products, which are discounted less through time than timber (see chapter 7). The relative contribution of non-timber goods would depend in large part on the extent to which they can be included without significantly decreasing timber production. The potential for doing this will be greater in the earlier stages of a plantation, when the timber trees are still relatively small and space is available for other plants, than toward the end of the plantation rotation. The morphology and resource requirements of the species in question are also important—in the case of vines (rattan, medicinals), herbs (condiments, medicinals) and shrubs (many fruits) the potential for competition with timber trees is minimized, whereas the inclusion of large trees would reduce the space available for timber trees. An exception to this latter point might include trees that produce both timber and non-timber products, such as *Durio* in Asia or *Brosimum* in Latin America. The inclusion of non-timber goods in forest plantations could also serve to direct a greater proportion of plantation benefits, via both products and employment, to local people. In densely populated areas, this aspect of multi-purpose plantations can hold tremendous implications not only for social equity, but for the often essential requirement of local cooperation in plantation protection as well.

Although often collected from natural forests or cultivated in small plots throughout the tropics, few so-called minor forest products currently play a significant role in commercial forest plantations. An important exception is found in Central and West Kalimantan, Indonesia, where the intercropping of rattan with rubber contributes to the incomes of plantation smallholders (Godoy 1988). The cultivation of rattan in a variety of forest settings has been tried experimentally in several Southeast Asian nations, and several multinational companies are contemplating the inclusion of rattan in large-scale forest plantations in Indonesia and Malaysia (Godoy 1988; see appendix 5.1). In Sri Lanka, heavily logged forest is undergoing enrichment planting with jackfruit (*Artocarpus*) for the production of both timber and fruit (P. Ashton, personal observation 1988).

The potential of non-timber products to increase the value of forest plantations is suggested by an economic analysis of alternative plantation designs in India (Bromley 1981). The study examines a number of model plantations developed for several sites in Madhya Pradesh, using hypothetical but carefully determined values for such parameters as plantation establishment and maintenance costs, land values, production rates, product prices, rural population densities, and local consumption patterns. Comprehensive cost benefit analyses of the model plantations are employed to determine the optimum mix of species and products under a variety of social and ecological conditions. A major finding of the study is the dramatic potential increase (up to 150 percent) in net present value made possible by diversifying fuelwood plantations to include trees that produce fruit, fodder, and construction materials (Bromley 1981). Since the analysis is directed toward community forestry development, particularly fuelwood plantations, timber production plays a relatively minor role in the models tested. It seems likely, however, that a mix of species and products would prove equally effective at increasing net present value in timber plantations grown in areas where population density was sufficient to fully utilize non-timber products.

Perhaps the best examples of the silvicultural potential and long-term stability of multiple-product tree plantations in the humid tropics are traditional home gardens. Found throughout the humid tropics, home gardens consist of relatively small (often less than one hectare), privately held plots of land on which a number of perennial plants, particularly trees, are planted and maintained in an orchard. Although highly variable from one region to another, they are characterized by a diversity of tree species as well as other life forms, mostly native, which produce a variety of products, including fruit, nuts, spices, latex, resins, medicines, fodder, fuelwood, and construction materials. Structurally, they resemble natural forests, with several layers of increasingly shade-tolerant trees and shrubs forming an overlapping network from the overstory to the ground. As in natural forests, light interception is often remarkably high in home gardens (Christanty et al. 1980), suggesting a photosynthetically efficient use of the available space.

Some of the most complex and carefully managed home gardens are found in west Java, where rural people have depended on them for centuries to provide products previously extracted from what are becoming increasingly scarce natural forests. Comprised of both perennial and annual plants, Javanese home gardens are highly diverse—in a study of 351 home gardens in western Java, 501 plant species were found to be cultivated (Karyono 1981). Home gardens are estimated to provide approximately 14 percent of the total carbohydrate and protein consumption of rural households (Soemarwoto and Soemarwoto 1984), although no distinction is made in this figure between perennial and annual crops. However, the contribution of home gardens to the rural economy, averaging approximately 25 percent of total household incomes, is derived almost entirely from tree products, specifically fruit and spices (Danoesastro 1980). In Sri Lanka, annual net incomes of up to $2,500/hectares/year have been predicted (from the twentieth year onward) for smallholder home gardens located on senescent tea plantations (Carpenter 1983).

In addition to planting home gardens near their dwellings, rural people can modify natural forest to resemble home garden systems through the selective removal of non-desired species. In Indonesia, this method is widely used to enhance the concentration of useful trees in village forests while leaving the stand structure essentially intact (Soemarwoto and Soemarwoto 1984). In addition to producing many of the products obtained from home gardens, these managed forest plots often provide a substantial portion of household fuel needs, mostly in the form of fallen branches and dead wood. Timber and poles from selectively harvested trees can be sold or used for domestic construction. Bamboo, often a dominant plant on more open sites, can also be harvested and sold as building material (Soemarwoto and Soemarwoto 1984).

Evidence of this type of forest management by the pre-colonial Maya of Mesoamerica is found in the form of isolated patches of tall trees surrounded by shorter forest in Yucatan, Mexico (Gomez-Pompa 1987). Most of these patches possess unusually high concentrations of useful plant species, many of which are commonly found in present-day Mayan home gardens throughout the region. In addition, some of the patches are surrounded by ancient stone walls, believed by the present-day Maya to have been built by the "old" Maya. These observations have led researchers to conclude that the patches represent remnants of the original primary forest which were subjected to selective thinning as a means of developing forest gardens. As such, they may constitute a transitional stage between the simple extraction of products from natural forest and Mayan home gardens as they exist today (Gomez-Pompa 1987).

Tropical home and forest orchards are oriented primarily toward the production of non-timber products for domestic or regional consumption. As currently designed, they hold limited potential for producing significant volumes of high-quality timber for the international market. However, on the basis of the limited data available, these garden systems offer the best models developed to date for using marginal lands for sustainable, intensive forest production under conditions of high population density and low capital availability. As such, they warrant careful study to determine how the salient features of their design might best be incorporated into timber-producing stands to increase both the economic profitability and social acceptability of forest plantations.

CONCLUSION

Forest plantations are widely believed to be an economically attractive and necessary means of enhancing wood production in the tropics. To date, the majority of these plantations consist of fast-growing species planted in monoculture. This plantation design, the most successful examples of which are found in the seasonally dry tropics, is currently being extended to the humid tropics on a large scale for the production of industrial wood and fuelwood. However, there is reason to question the long-term sustainability of short-rotation monoculture plantations in the humid tropics on the grounds of increased susceptibility to insect and disease attacks, deterioration of soil nutrient status and structure, and the inefficient use of growing space. In addition, the widespread practice of developing forest plantations through the conversion of natural forest ignores the loss of non-timber products and forest services, and may actually reduce the overall production of timber as well.

Mixed-species plantations present an alternative means of producing tropical timber on marginal lands. In particular, the establishment of mature-phase trees beneath a cover crop of fast-growing, building-phase trees offers the possibility of producing both industrial wood and high-quality timber on the same site by emulating the successional processes found in natural forests. One method of accomplishing this might be through intensive enrichment planting in heavily-logged or otherwise degraded forests. Mixed-species plantations could also be created on deforested lands, using nitrogen-fixing and/or deep-rooted trees to facilitate soil restoration. In either case, plantations should not be placed on sites with existing productive natural forest.

The inclusion of non-timber products such as cane, fruit, spices, and fuelwood in forest plantations holds promise for (1) increasing the net present value of timber plantations by providing an early and continuing flow of market products, (2) providing employment in areas of high population density, and (3) increasing the distributional equity of plantation benefits by producing locally needed goods and services. The viability of mixed-product plantations will depend on a number of social and environmental factors, including the availability of labor, local need for non-timber products, ability to equitably allocate the costs and benefits of plantations among the interested parties (government, private corporations, and local communities), and the degree of ecological competition between timber and non-timber species. To date, the best available models of multiple-use plantations are traditional home and forest gardens. Additional research, including a rigorous evaluation of existing models and field trials of new designs, will be required to adequately assess the ability of multiple-use plantations to produce timber in a cost-effective and sustainable manner.

CHAPTER 10

Conservation of Genetic Resources

GENETIC RESOURCES, both of plants and of animals, provide the necessary means both for the continued improvement of existing crops and for the development of new crops in response to ongoing changes in demand and technological progress. Genetic resources, manifest in the diversity of plants and animals in nature, are also of increasing interest to the public worldwide (especially in developed countries), and have become a major resource for the tourism industry.

Between two-thirds and three-quarters of all biological diversity is harbored in the world's tropical evergreen forests (Raven 1980). The major reserve of genetic diversity necessary for development (and biological protection) of future agricultural plantation and forest crops for the humid tropics resides in the natural forests. While human-made plantations maintain rainfall regimes and cycle carbon at a rate similar to that of indigenous forests, genetic resources are eliminated when indigenous forest is converted to plantations. Therefore, conservation of genetic resources, through adequate conservation of indigenous biological diversity, must be evaluated with other goods and services provided by forests, and must form an integral part of land use and management.

THE CHARACTERISTICS OF GENETIC RESOURCES

The choice of sites for genetic resource conservation, and the determination of their size, should be based on the biological characteristics of the species whose genetic resources are to be conserved. Species vary

190

enormously in their intrinsic genetic variability. Variation within individual populations may or may not exceed regional variation. Their variation is influenced by breadth of ecological range, breeding system, and dispersal characteristics (Ledig 1986). Small populations are more prone than large to chance destruction as a consequence of unexpected catastrophes such as drought or disease epidemic. Also, fragmentation of species populations into permanent small isolates both promotes opportunities for genetic differentiation among them (Mayr 1963) and reduces opportunities for the spread of genes.

Any reduction in area inevitably leads, in time, to some extinction and thereby to some attrition of genetic diversity (Macarthur and Wilson 1967). The aim of conservation, therefore, must be to optimize genetic diversity along with other goods provided by the forest, rather than to preserve everything, which in practice is impossible. Species and genetic diversity are optimized in large unitary areas and where corridors of suitable habitat exist that will permit migration between the forests (Diamond and Case 1986; Forman and Godran 1986; Lovejoy et al. 1986). Enough is now known of patterns of species diversity in tropical forests, and of the requirements of different species, to prescribe general rules for conservation of genetic resources within them.

All animals ultimately depend on plants for food, but most plants in rain forests depend on animals for dispersal of pollen and seeds. It is now thought that the principal determinant of overall species diversity in tropical forest ecosystems is the level of plant species diversity. This is because the major portion of biological diversity is composed of insects. Plants, to defend themselves against insect herbivores, have evolved specific chemical defenses (Janzen 1971). As a consequence, each plant species hosts specific insects that have evolved means to overcome its particular chemical defense. Therefore, by allowing more plant species to survive, more insect species will also be preserved (T. Erwin, personal communication).

These facts are central to understanding the importance of genetic resources of tropical forest, and how to maintain them. Plant chemical defenses are manifested as forest chemical products (see chapter 6) including latexes, pharmaceutically active compounds, and compounds that can be used in cosmetics. Also, competition between plants and species-specific herbivores and pathogens can mediate plant population densities. The equable climate of the humid tropics favors plant pests and diseases, and this prevents plants from surviving long in single-species stands in nature. On this account, their low sustainable population densities allow space for other plant species to survive, resulting in a composite of numerous species in the same area. This is a

major reason for high biological diversity in tropical rain forests (Janzen 1971; Hubbell and Foster 1987). It is also the major reason why
multiple-species plantations are advocated as opposed to single-species
plantations (see chapter 9).

By the same token, rain forest plants and animals form symbiotic
interdependencies, which are often highly specific, with other organisms. Examples include the dependence of most tropical plants on
animals for pollen and seed dispersal, while many animals are dependent on pollen, nectar, fruit, or seeds as food; likewise, most plants
depend on fungal mycorrhiza for nutrient intake, while the fungi rely
on autotrophic plants for carbohydrates. This underlines the necessity
of conserving biological diversity, hence genetic resources, *in nature*.
Botanical and zoological gardens are invaluable tools for research,
particularly for studying the biological characteristics whose understanding is essential for conservation management of natural populations. But it is generally impractical and uneconomic to conserve
population samples of rain forest organisms in gardens, plantations, or
zoos, which might adequately represent species genetic diversity, but
cannot maintain many of the symbiotic relationships on which species
populations depend for survival. Knowledge is currently inadequate to
maintain some symbiotic interdependencies *ex situ*, while in other
cases, such as that of many species of specialized pollinators that
require alternative food sources year-round, maintaining symbiotic
interdependencies may be impractical *in situ*. P. Ashton (unpublished,
1986) calculated the cost of maintenance of a population sample of a
tree species large enough for conservation of genetic diversity to be
about $500 per annum in the United States, which makes *ex situ*
conservation difficult to justify, except for species of economic importance.

In the present context, these facts indicate that knowledge of patterns of plant species diversity provides the most ready and accurate
approximation to overall species diversity, critical to conservation planning. Plant species diversity (that is, the number of species) within
tropical forests varies geographically and with climate, degree of disturbance, and soil fertility. In tropical forests, plant species diversity
increases broadly with increases in annual rainfall (Gentry 1988) and
with decreases in the length of the annual dry season (Ashton, in
prep.). Species diversity is greatest in the lowlands and declines with
altitude, except where rainfall seasonality decreases with altitude. Diversity is greatest on soils of moderately low, but not very low, fertility
(Ashton 1977; Ashton, in prep.; Huston 1978; Tilman 1982). Thus,
Ashton found that species richness in lowland Borneo forests peaked

on leached humic ultisols on sandstones and rhyolites, but declined both on infertile spodosols and on udult ultisols over shales or basalt. But tropical plants often occupy narrow ranges of soil fertility, so that forests on each part of the soil fertility range require conservation if their distinctive floras are to be adequately represented in conservation (Ashton, in prep.). Further, vertebrates, including species important in seed dispersal, may depend on fertile valley and other sites within an overall infertile landscape for maintenance of their populations. Therefore, conservation of diverse habitats that may be of marginal value alone is not sufficient. Rather, such habitats must be conserved within more extensive landscapes that include at least some fertile sites.

Plant species diversity is optimal at moderate levels of climatic disturbance, allowing persistence of large areas of the mature phase of the forest, and the characteristic species it contains, as well as the less rich regenerative phase of the forest. Selective logging, as currently practiced, does not simulate these conditions, because it reverses this relationship, causing the mature phase instead to be fragmented into islands with consequent dramatic increase in the likelihood of local extinction through island effects (Macarthur and Wilson 1967). This may also reduce frequency of cross-pollination, which can reduce levels of seed production and may eventually lead to extinction (Hamrick and Murawski 1990). Extinction rates will accelerate in subsequent felling cycles, when the islands will be replaced by others that have regenerated from areas logged previously. The quantitative change in relative area of forest successional stages will, in time and after successive felling cycles, lead to lasting changes in population densities of plant species. Light demanders will become relatively more abundant, whereas the shade-tolerant majority will decline. Theoretical models predict that many of these, already rare in primary forest, will disappear (S. Pacala, personal communication). Also, valued slow-growing species may be harvested too frequently for a sufficient number of individuals to reach reproductive maturity within a felling cycle. Logging, even when selective, and conservation of genetic resources are therefore incompatible.

Species diversity varies from region to region due to isolation and diversity of sites, immigration of new species and extinction of existing ones (Macarthur and Wilson 1967), or differing histories of environmental changes or catastrophe.

Plant species diversity is greatest on a regional scale in the neotropics, where about 90,000 species are thought to occur, compared with 45,000 in tropical Asia and Australasia, and 35,000 in Africa and Madagascar (Raven 1980). Specifically, the foothills of the equatorial and northern Andes host the most species-rich of all plant communities (Gentry 1988). The lower montane cloud forests of the equatorial and northern Andes are renowned for their exceptionally rich epiphyte and shrub floras. Tree species richness within plant communities reaches its highest levels more or less equally in the Andean foothills and the forests of northwest Borneo (Ashton, in prep.). The richest African forests are in Cameroon, and possibly along the northeast coast of Madagascar where endemism is also exceptional. Forest communities in the foothills of southern and western New Guinea and the lowland forests of peninsular Malaysia are also exceptionally rich by world standards.

Plant species, unlike most animals, vary greatly in their breeding systems, and these are major determinants of genetic variability. Overwhelmingly, rain forest plants are outbreeders, though often facultatively so (Bawa 1976). Most pollen and seed dispersal is over short distances (Ridley 1930). Species examined show a greater range of genetic variability within breeding populations than exists between populations, even if these populations are great distances apart (Gan et al. 1981; Hamrick 1983; Ledig 1986). A significant number, still to be ascertained, produce genetically identical offspring through asexual seed production (adventive embryony). This reduces population genetic variability, but appears to be associated with high regional variability within species, particularly of species occupying specialized habitats (Ashton 1984).

The minimum viable size of a plant species population depends on its breeding system, pattern of genetic variability, and susceptibility to sudden fluctuations in numbers. On theoretical grounds, it has been asserted that 200 reproductive individuals constitute the minimum number for a viable population of outbreeding trees (Whitmore 1977). Even in the most diverse tropical rainforests, 5,000 hectares of each habitat will achieve this goal for 95 percent of species (Ashton 1984). Up to now, conservation organizations have used vertebrates, particularly large mammals and birds, as leading indicators in the setting of priorities for conservation of areas of tropical forest. These animals are mostly generalized feeders, and the few specialists are not representative of biological diversity as a whole. Nevertheless, many rain forest trees depend on such animals for seed dispersal, and sometimes successful germination. Some generalized feeders, particularly larger browsing animals such as elephants and rhinoceroses, can require as much as 100,000 hectares for conservation of stable populations (Caldecott 1987). But they are generally less affected by selective logging than specialists. A system of strict preserves, set in a buffer zone and connected by corridors of natural forest managed for timber production, will therefore satisfactorily serve the needs of most generalists (Forman and Godran 1986). This concept of allocating blocks within a multiple-use forested landscape for strict preservation has been advocated by Bowes and Krutilla (1989). Care must be taken to ensure that loggers do not illegally invade preserves, as has happened in the virgin jungle reserves set within production forest reserves in peninsular Malaysia. (See table 10.1 for examples of management systems to maintain biological diversity.)

The capacity of natural production forests to serve as corridors in order to assist conservation of wide-ranging animal species is much enhanced by conservation of certain keystone plant species. Examples include populations of strangling figs and certain understory tree genera, which provide fruit continuously and are thus critical during times of scarcity. This can be easily achieved without detriment to production of timber (or other goods) by leaving them in the residual stand. Indeed, the proximity of forest blocks with differing uses that complement one another in this way adds to the total value of the forest (Bowes and Krutilla 1989).

METHODS OF SITE EVALUATION FOR SELECTION OF STRICT PRESERVES FOR CONSERVATION OF GENETIC DIVERSITY

Patterns of regional concentration of species richness and endemism form the primary basis for establishment of a strategy for conservation of genetic resources. Direct evidence is derived from individual species distributions, with particular attention paid to ecologically specialized species. In regions where knowledge of biota remains limited, geological maps (and specifically maps of surface lithology) provide a satisfactory guide, if regional climate and the relationship between surface geology and soil are broadly understood.

Fortunately, tree species are almost entirely identifiable by experienced field biologists on the basis of their fallen leaves. This provides a rapid means of surveying individual forests for the setting of priorities for conservation. It would be quite feasible, through such organizations as the Botanic Gardens Conservation Secretariat of the International Union for the Conservation of Nature and Natural Resources (IUCN), to train specialists at the country level to undertake such work, and to provide specialists from the international community of taxonomic botanists in the interim.

In general, 10 percent of the tree species in a tropical forest comprise

Table 10.1

EXAMPLES OF MANAGEMENT SYSTEMS TO MAINTAIN BIOLOGICAL DIVERSITY

Onsite		Offsite	
Ecosystem Maintenance	*Species Management*	*Living Collections*	*Germplasm Storage*
National Parks	Agroecosystems	Zoological Parks	Seed and Pollen Banks
Research Areas	Wildlife Refuges	Botanic Gardens	Semen, Ova, and Embryo Banks
Marine Sanctuaries	In-situ Genebanks	Field Collections	Microbial Cultures
Resource Development	Game Parks and Reserves	Captive Breeding Programs	Tissue Culture
	Planning		

Increasing Human Intervention ⟶   ⟵ Increasing Natural Processes

BASED ON: World Bank 1987.

at least 50 percent of the stand. Endemic and ecologically specialized species are as well represented among the commonest 10 percent as the rarer 90 percent (Ashton, in prep.). Rapid extensive species surveys, which provide reliable information concerning the commonest 10 percent of species only, can therefore give information on conservation priorities not only from the point of view of species richness but also of the specific genetic resources of a forest.

CONCLUSION

Specialized animals and plants can survive in relatively small areas that must nevertheless be maintained in a completely unexploited, unmodified state. Generalized species often require larger areas, but cyclical selective exploitation is not generally harmful to them. The minimal area for each habitat in a strict preserve should, where possible, be 5,000 hectares (though it is recognized that there are exceptions in nature, such as mountain peaks and limestone outcrops). A few large, environmentally heterogeneous preserves are preferable to many small, environmentally uniform preserves. Small, strict preserves should, where possible, at least be connected by corridors of natural production forest, and at best sit in an extensive matrix of it, the whole comprising at least 100,000 hectares. Keystone food plants, such as figs, should be conserved in such managed production forests. The full range of sites, in each climatic zone, needs to have adequate representation for conservation of genetic resources. Particularly threatened are the genetic resources of arable lands.

Sites of highest priority for genetic conservation are those of high species endemism (species unique to the site) and high species richness. High endemism is likely to be observed on sites containing specialized habitats, or an unbroken history of uniform conditions. High species richness is likely to be present on sites of equable humid climate (i.e., evenly distributed high rainfall, or, in the case of epiphytes, constant mistiness), areas of moderately low soil fertility, and locations with moderate levels of natural disturbance. High vulnerability to degradation or deforestation must also be recognized as a criterion for evaluating priority.

CHAPTER 11

Institutional Constraints and Options: Customary Rights vs. State Ownership

THE PAST mismanagement (or rather absence of management) of tropical forests has its roots in the prevailing institutional arrangements that determine who owns and controls the use of the forest, and who has access to and uses the products of the forest. A forest that is not securely and exclusively owned by an entity—a state, community, or an individual—that can effectively enforce its ownership rights to the exclusion of all others is in effect common property and is bound to suffer the "tragedy of the commons" (Hardin 1968). Rather than being managed, "common-property" or open-access forests are frequently mined, degraded, and ultimately abandoned.* The lack of interest in investing in the enhancement and management of an open-access resource is traceable to the lack of assurance that the benefits from this investment will ever materialize. Investments that would otherwise be very attractive will not be made if the return is to be shared among an unspecified number of claimants who have free

---

* "Common property" is used here in the sense of open-access rather than communal property. Communal property or property used under customary rights are known to be effective forms of resource ownership in well-organized and socially cohesive communities.

access without incurring any part of the investment costs. In developing this argument, nominal ownership and legal status must be distinguished from effective ownership and actual use.

Historically, most tropical forests have been communal or tribal domain to which the members of the community or tribe have had customary rights of access and use. During the past 150 years, over 80 percent of the world's tropical forests have been brought under government ownership, and forest legislation has been introduced to give special legal status to particular areas of the forests such as forest reserves, protected forests, gazetted forests, and national parks. However, governments throughout the tropics have been relatively powerless to enforce their ownership rights and defend the legal status of the forest for several reasons, including:

1. the vastness of the areas transferred to state ownership (in most countries over 50 percent of total land area);
2. the speed and manner in which the transfer of ownership has been made;
3. the failure to recognize and accommodate the customary rights of individuals and communities to the forest, which has created resentment among local populations;
4. the limited budget and administrative, technical, and enforcement capacities of the newly established estates;
5. growing pressures from expanding rural populations; and
6. the failure of rural development to provide alternative employment and income opportunities.

As a result of governments' inability to uphold their ownership and enforce related forest laws, tropical forests have reverted to quasi-open access with pervasive encroachment, squatting, log poaching, slash-and-burn (shifting cultivation), and illegal forest conversion to other uses. At the same time, the governments award concessions to logging companies on truly concessionary terms and fail to adequately enforce harvesting and replanting regulations. Logging companies operating under short-term concession agreements, without assurance of renewal and with constant threats of encroachment and poaching, adopt hit-and-run strategies, since they have no incentive to preserve the long-term productivity of their concessions. The result is a pervasive climate of lawlessness, uncertainty, and insecurity of tenure for all parties (government, logging companies, squatters, and local communities), none of which have sufficient control and incentive to conserve the resource base or invest in its proper management and enhancement.

Most tropical forests are *de jure* state property, but *de facto* open access with an undefined but large number of nonexclusive claimants. The government is in the tropical forest sector primarily to earn tax revenues and foreign exchange from timber, but also presumably to assure environmental forest regeneration, watershed protection and other environmental services, and sustainment of traditional social values. The logging companies are in the tropical forest exclusively for profits from timber harvesting. Local populations are in the forest for fodder and fuelwood collection, crop cultivation, and gathering of a variety of non-timber goods such as fruits, vegetables, meat, fish, fibers, and medicinal products primarily for domestic use but also for sale.

In principle, these are not incompatible interests. Tropical forests can be managed for multiple uses, as we have seen in chapter 8, but such management is all but impossible without clearly defined property rights and security of tenure. Institutional reform is a necessary condition for multiple-use management of tropical forests and investment in their regeneration and enhancement, insofar as it addresses these issues of ownership, legal status, and insecurity of tenure.

OWNERSHIP, LEGAL STATUS, AND MANAGEMENT OF TROPICAL FORESTS

Although the three tropical regions of Africa, Asia, and Latin America have had different traditions of ownership and access to forests, they all have experienced the same postwar trend toward state ownership. Tropical Africa has had a tradition of customary or usage rights related to hunting, grazing, shifting cultivation, and gathering of forest products such as building materials, fuelwood, vegetables, and animal products. Traditionally, forests were considered the property of communities or tribes and managed by village headmen or tribal chiefs who were "responsible for the implementation of limiting principles or rights by families giving full respect to collective interests and to the conservation of natural resources for future generations" (Lanly 1982).

The British and French colonial governments introduced written forest laws based on their respective traditions at home. In many cases, British administrations recognized the claims of local populations to forest ownership and incorporated much of customary law in the codified forest laws. Ghana, Nigeria, and Malawi are good examples of former British colonies where the forests were declared the property of traditional communities. State forests were created only with the consent of the community and often without altering their ownership status (Lanly 1982).

In contrast, Roman law, on which the forest laws introduced by the French administration were based, declares all unoccupied lands without written ownership documents as state property, thereby contradicting the nonliterate customary law of local usage rights. Following independence, there has been a general tendency to declare forest to be state property throughout Africa, regardless of local traditions, colonial history, or socioeconomic system. Of the six African countries studied, Gabon and Ivory Coast effectively maintained 100 percent of the forests under state ownership even though they have created community forests on paper.

In Ghana all forest lands, which were owned by traditional communities until independence and held in trust for them since then by the central government, have been transferred entirely to central government ownership since 1973. Thus, Ghana has moved practically overnight from 100 percent communal ownership to 100 percent state ownership. In Liberia, all tropical forest areas have been declared national forest under state ownership, and shifting cultivation is prohibited in national forest areas (FAO 1981). In the Congo, the forest law that provided for private communal forests was abolished in 1973 and forests were declared state property. Only Cameroon's 1974 Forest Law explicitly recognizes, in addition to state forests, public community forests, private forests, and forests managed by the state in which local populations have usage rights.

The tropical forests of Africa, therefore, despite their long tradition of communal ownership and customary usage rights, have been brought increasingly under central government ownership. Private ownership of forest land is virtually absent in tropical Africa with the possible exception of Cameroon and South Africa.

Tropical Asia has had a tradition of both private and communal forest ownership. Before independence in South Asia, particularly India, many forests were owned by princely estates, private individuals, and religious societies, as well as communities, while in Southeast Asia (Indonesia, Malaysia, Papua New Guinea) many forests were recognized domains of communities and clans or tribes. Today, 80 to 90 percent of the forest area in the region is government owned and is managed by forest departments (Lanly 1982). The process of nationalization of forestlands in Asia began earlier and was more gradual than in Africa. In India private ownership has been gradually abolished, and in Indonesia the government assumed all property rights to the natural forest through provisions in the 1946 Constitution (Repetto and Gillis 1988). In Malaysia the property rights over the natural forests have been assumed by the state governments. In both Thailand and the Philippines virtually all natural forests are under government ownership. In Thailand, which was never a Western colony, forests belonged to the feudal chiefs until 1890 when the forest ownership was transferred to the government (TDRI 1987). However, despite the marked increase in direct forest ownership by the state, there are still limited forest holdings by corporations, communities, trusts, and temples. Yet, by far the dominant actor is the Forestry Department, which in many countries (India, Thailand, Indonesia, Malaysia) shares administrative and management responsibilities with state forest enterprises.

The most notable exception to the general increase of state ownership and control is Papua New Guinea, where forests are not owned by the government but by clans and tribes that are recognized as cooperatives and can deal directly with the logging companies. The government must negotiate with clans and tribes for the right to use forest resources and has already purchased timber rights on over 2 million hectares from the communities (Lanly 1982). To summarize, state ownership of forests is almost as pervasive in Asia as it is in Africa, but both communal and private property are somewhat more prevalent.

Over 80 percent of the Latin American tropical forests is under state ownership; the balance is held as private property (Lanly 1982). State ownership accounts for 100 percent of forests in Peru, almost 100 percent of forests in Bolivia, and over 80 percent in Brazil and in Trinidad and Tobago. It is somewhat less significant in Ecuador and Honduras. Private ownership is relatively important, especially in Central America, Mexico, and Brazil, where coniferous forests are generally held privately. Communal ownership of forests is not significant in Latin America, with the exception of Mexico, where 50 percent of all forest area are communal ejidal lands.

Thus, in all three regions the state is by far the principal forest owner with private ownership ranging from a high of about 20 percent in Latin America to being virtually nonexistent in Africa. Customary usage rights have predominated in Africa, and communal ownership has been important traditionally in Asia; both continue to exist in a more limited and weakened form in some countries today.

While state ownership of forest resources has been growing in all regions and is currently dominant in almost all tropical countries, the legal status of forests varies across regions and even among individual countries. In Latin America, a distinction is drawn between "production forests," which are reserved for the production of wood, "protection forests," retained for land and water conservation to the exclusion of logging, and national parks or "nature reserves," which are set aside for a variety of functions including recreation, wildlife refuge, and preservation of biological diversity. The total area closed to logging in

Latin America is 16 million hectares or about 1.8 percent of all forests (Lanly 1982, 54).

In Africa, a distinction is usually made among gazetted forests, protected forests, and national parks, but the legal content of each category differs according to colonial background. In French and Belgian colonies, gazetted forests could not be accessed by persons or communities without prior degazetting: agriculture was strictly prohibited, and logging was regulated. Where the law was not strictly enforced, gazetted forests were encroached upon and degazetting often followed (e.g., Ivory Coast). In contrast, in the British colonies, persons and communities with prior rights to the gazetted reserves were allowed to maintain certain uses of the forests, including felling for local products, a use broadly permitted in unreserved forests. "Protection" of forests in French colonies meant that unauthorized logging or clearing for agriculture was prohibited.

Since independence, several African countries have amended their forest laws in an attempt to create permanent forest reserves, but only Liberia has succeeded in establishing national forests that cover 80 percent of existing forests and are relatively well protected from encroachment. Africa as a region has set aside as national parks and equivalent reserves 7.3 percent of its forested areas, or 51 million hectares of tree formations (of which 42 million hectares are mixed forest-grasslands).

In tropical Asia, major legal distinctions are drawn among forest reserves, protected forests, unclassified forests, and national parks, although exceptions exist in individual countries. In forest reserves controlled by forestry departments, unauthorized felling, collection of fuelwood and other forest products, grazing, and trespassing are prohibited. Forest reserves account for about one-half the forest lands of India, 37 percent of the forest of Sabah, and 8 percent of the forest area of Sarawak. Thailand has reserved 15 million hectares of forest (Lanly 1982). "Protected forests" are similar to forest reserves but much less control is exercised and villagers are generally allowed to graze their animals and collect small timber, fuelwood, fodder, and other forest products for their own consumption. Unclassified forests are, like the other two classes, state property, but their legal status is not yet defined. The major exceptions are the Philippines, where forest and alienable lands are distinguished, and Indonesia, which has not defined the legal status of its forests other than by prescribing the exercise of substantial government control over public forests and private plantations. Tropical Asia has set aside a total of 18 million hectares (mostly closed broadleaved forest), equal to 5.4 percent of the total forest area, as national parks, wildlife sanctuaries, and biosphere reserves.

Altogether, a total of 85 million hectares of forest area (or 4.45 percent of the total forest area of the three regions) has been set aside for conservation purposes (Lanly 1982). With the exception of parts of Latin America, little forest has been set aside for protective purposes (water and soil conservation) or for the use of local communities. The primary purpose of forest laws has been to establish state property and restrict the access of individuals and communities to the forests. Villagers in the vicinity of the forest are legally prohibited from collecting fuelwood and other non-timber goods, much less harvesting timber from "forest reserves," "gazetted forests," or even "production forests." However, regardless of the legal status of forests, encroachment is often widespread because of both enforcement problems and the lack of alternatives.

Whatever their legal status, tropical forests that are largely government owned (80 to 90 percent) are rarely managed. The principal exception is national parks and equivalent reserves (4.4 percent of all forest), where logging is prohibited and intensive management is practiced for the development of recreation, tourism, wildlife, and general conservation. Using the FAO's definition of intensive management to include control and regulation of logging, protection from fire and disease, and silvicultural treatments, the forests of tropical America generally are not intensively managed. Exceptions include small, managed areas in Trinidad and Tobago and Honduras, and silvicultural trials in Brazil and Peru, which have prepared, but not yet implemented, management plans.

Several African countries including Ghana and Zaire have long had management plans that provided for both harvesting regulations and silvicultural treatments, but, except for a brief period in the early 1960s, intensive management has rarely been implemented due to the lack of resources and increasing population pressure. Ghana continues to manage as much as 90 percent of its productive closed forest (1.2 million hectares) and thus accounts for two-thirds of all intensively managed forests of Africa. Uganda (0.44 million hectares) and Kenya (0.07 million hectares) account for most of the remaining third; Cameroon and the Ivory Coast are presently experimenting with intensive management systems.

In total, less than 2 million hectares of productive forests are being managed out of Africa's total of 164 million hectares. However, several English-speaking countries of Africa practice some form of extensive management. Among the rest, most notable is the case of Congo, cited by Lanly (1982), whose forest law provides for "elaboration of management plans including not only forest logging but also tourism, hunting rights, wildlife protection and, as far as forest protection is concerned, the determination of a maximum annual exploitable volume (VMA) for the main commercial species, infrastructure planning and silvicultural prescriptions."

Tropical Asia, especially in former British colonies, has the longest history of intensive forest management, particularly on the Indian subcontinent and Malaysia. More is known about silviculture and management in South Asia, where most of the managed forests are deciduous, than in Southeast Asia, where evergreens and semideciduous forests dominate and have more complex ecosystems (see chapter 3). Two systems of management have been used in Asia: (1) the monocyclic (or shelterwood) system, which aims at obtaining uniform crops for subsequent harvest, and (2) the polycyclic system, which involves selective felling with or without silvicultural treatment (see chapter 3 for details). The Malayan Uniform System (of the monocyclic variety) was practiced in the 1960s in the lowland forests of Malaysia but has since been replaced by polycyclic management systems. The latter are currently common in Southeast Asia in theory but often, and perhaps always, fail to foster sufficient regeneration because of lack of adequate management, encroachment by shifting cultivators and squatters, or inherent failure to achieve expected stocking or growth rates.

GENERALIZED INSECURITY OF TENURE

The sudden transfer of forest resources to government ownership after centuries of community and private ownership and use, when done without due consideration of customary rights, has created a climate of uncertainty and bias against forest investments and a pervasive distrust of government forest policies and projects. For example, Cernea (1981) reports that farmers in Pakistan have been reluctant to accept government tree planting for fuelwood on their lands; he offers this explanation:

> The hesitations of smaller farmers stem from their suspicions (1) that they may lose possession or control over their land to the government once it has been planted with trees by the Forest Department, and (2) that they may be deprived of access to fodder collection and grazing which is critical for them. Most of the smaller farmers interviewed indicated that they would, if they could, offer small plots for project planting provided they receive convincing assurances that the Forest Department will not alienate

their lands and they will be able to cut grass for their cattle.
(Cernea 1981, 21)

While relatively successful traditional systems of forest management
have been dismantled in favor of centralized management, the perceived benefits did not follow because governments have been unable
to put in place alternative management systems that could be effectively implemented from the capital with limited budgetary and administrative resources. Forest laws remained largely on paper and even
the most basic forms of management—protection from encroachment
and enforcement of logging agreements and regulations—have not
been exercised. Valuable forest resources have been leased out for
logging at truly concessionary terms with little regard for the conservation of the resource base.

A well-intentioned attempt to assert state ownership by limiting the
length of the concessions can have devastating effects on the resource
base and forest regeneration after logging. As documented in chapter
4, the span of most logging concessions is under 25 years and in many
cases under 10 years, while it takes 50 to 70 years to grow tropical
hardwoods. Short-term concessions lead to short-lived forests. In Sabah
and Indonesia between 45 and 74 percent of trees that remain after
harvesting operations are substantially damaged or destroyed (Repetto
1986; see also chapter 8). A climate of uncertainty is created by the
short span of the concession. The concessionaire is uncertain whether
any investment made in conservation or replanting will be recouped,
since the concession agreement runs out before the next harvest and
there is no assurance of its renewal. Thus short concessions coupled
with inappropriate tax structures encourage a selective "cut and run"
strategy (high grading), with little concern for the remaining stand and
future productivity. Moreover, according to Repetto (1986):

> Political instability and pressures from local "partners," irregular-
> ities in the contracting process, and risks that one-sided agreements
> will be reexamined and renegotiated, all lead concessionaires to
> realize their profits as early as possible. (Repetto 1986, 68)

Hard-pressed individuals and communities that have lost their customary rights to forest resources without commensurate alternatives
quickly perceive the inability of the government to enforce its ownership over the resource. They encroach on logged-over forests opened
up by logging roads, slash and burn the remaining stand, and practice
shifting cultivation or sedentary agriculture. Encroachment and squatting on public forest lands and concessions create further uncertainty

for both the government and the concessionaire, inducing the former to further undervalue the forest resource and the latter to adopt an even more short-term harvesting strategy.

Understandably, governments unable or unwilling to expel encroachers and squatters from public lands, because of socioeconomic and population pressures, assert their ownership of the resource by refusing to recognize any legal rights of squatters to the land they occupy and by continuing to consider the encroached areas as public forestlands. This compounds uncertainty and damages the resource base even further, because without a secure title to the land, squatters are both unwilling (no incentive) and unable (no access to credit) to invest in land improvement or tree crops that are often more suitable than annual crops to the soil and water conditions of the land in question.

Thailand provides a good example of this situation, which is typical of many tropical countries, including those that are major timber producers. According to a World Bank study by Feder, Gershon, Onchan, and Hongladarom (1986):

> [In Thailand] large numbers of farmers do not have legal ownership of the land which they operate, even though they are perceived as *de facto* owners within the farming community. This situation characterizes in particular about one million households of squatters settled on lands officially designated as forest reserve lands. (Feder et al. 1986, i)

The authors of the above study conducted detailed surveys and rigorous econometric analyses. They have found a number of important results regarding the effect of insecurity of tenure on investment, productivity, and land values: (1) the value of squatters' land was only one-half to two-thirds the value of legally owned comparable land; (2) the capital/land ratios in securely owned lands were 56 to 253 percent higher than in squatters' land; and (3) the crop value per unit of land in securely titled land was 12 to 26 percent higher. Finally, the authors found that the social benefits from providing land ownership security range between 25 and 80 percent of the market value of squatters' land, and that the private benefits to the farmers are even higher (Feder et al. 1986, 14). These are enormous benefits, if one considers that over 40 percent of the agricultural land in Thailand is occupied without a secure title; it would amount to roughly 10 to 30 percent increase in Thailand's total agricultural productivity.

Many governments do recognize these problems but are reluctant to issue ownership titles to squatters for fear that they will encourage further encroachment and land grabbing. This is a legitimate concern but the problem cannot be solved either by ignoring it or by issuing usufruct (right to use) certificates to squatters, an option tested in Thailand through the *STR* (Sor Tor Kor, or land-use permit) program (TDRI 1987). Such certificates do not confer full property rights and thus preclude the ability to sell or mortgage the land to obtain credit. Feder et al. (1986, iii) found that in the case of Thailand the provision of usufruct certificates neither encouraged land improvements nor raised productivity and incomes to the same extent as the provision of secure titles of complete ownership.

The continued threat of encroachment and squatting in Thailand (and elsewhere) can only be stemmed by the government reducing its ownership of public lands (in Thailand at present, 62 percent of the country's total land mass) to manageable proportions. Lands remaining public should include critical watersheds, fragile lands, national parks, and wildlife and nature reserves. Effective enforcement of state ownership is close to impossible as long as 26 million hectares, 51 percent of Thailand's total land area, is officially considered public forest land; Thai forests have already shrunk to 29 percent of that area (TDRI 1987).

INSTITUTIONAL REFORMS

Several policy changes and measures to improve the status and prospects of tropical forests are being taken or considered by governments throughout the tropics. These include, among others, "regulations restricting forest clearance, stoppage of encroachment through greater vigilance, . . . greater degree of control over logging operations, intensive management in the postharvest period, increased scale of reforestation, agroforestry, silviculture [and] fuelwood plantations" (FAO 1981, 107). However, experience suggests that these measures provide no more than temporary relief. No matter how many new regulations are enacted or technological improvements introduced, no matter how many trees ("miracle" or otherwise) are planted or intensively managed, the process of forest depletion still seems irreversible.

The problem is neither legal nor technical, but institutional and socioeconomic. State ownership of forests in the context of high population pressure and low enforcement capacity effectively means open access, which leads inescapably to the "tragedy of the commons" (i.e., physical degradation and dissipation of economic rent).

The acceleration of deforestation and degradation in the course of the past 30 years is not unrelated to the tremendous and concurrent increase in both population pressure and state ownership of resources. The larger the number of rural poor and the more extensive (and ineffective) the state ownership of resources, the more the situation approaches that of open access, in which the rule of capture is the only rational management rule. Moreover, the insecure tenure of loggers/concessionaires, shifting cultivators, and squatters induces exploitative behavior for short-term profit at the expense of long-term harvest. Were the forests securely owned by parties that could effectively enforce their ownership and control to the exclusion of non-owners, the problems of forests could be addressed and less excessive deforestation would have occurred (assuming, of course, no refractory externalities that cannot be internalized and no discrepancy between private and social rates of discount).

Yet only the problems of the trees would be solved under this scenario, not the problems of the people who depend on the forest for livelihood, unless income transfers happen to work, which is relatively unusual in the developing world. Forest depletion is primarily a human problem. The rural poor continue to be in dire need of fuelwood and fodder, and fundamentally lack resources to maintain a decent livelihood. They would still have few means to satisfy those needs or even express them except by theft, starvation, or revolution.

At present, the "open-access" state-owned forests are no more than a safety valve to ease the socioeconomic and political tensions that build up when increasing numbers of people find themselves without a source of livelihood. Forests and other natural resources, nominally under state control but effectively unappropriated except by capture, have long been resources of last resort for the rural poor. In the absence of such forests, rural-urban migration and undisguised unemployment, landlessness, and starvation would have assumed far greater proportions than they already have.

Thus, the root causes of excessive deforestation and resource degradation have been the open-access status and insecure tenure of these resources and the gross inequality in the distribution of opportunities and resources in general. However, as there cannot be absolutely exclusive and secure property rights (especially in the presence of externalities), there cannot be absolute equality. But a reasonable dose of both secure property rights and equality will go a long way in saving both people and forests.

While a more equitable distribution of all resources is called for, forest resources are a good starting place. Being legal property of the state and effectively open access, these resources are relatively free of the thorny problems raised by land reform and asset redistribution. There are many people desperate to own any productive resources and many forests equally "desperate" to have secure, effective owners.

Of course, not all forestland can be disposed of in this manner. A most important exception is watershed areas with significant externalities (downstream effects) that cannot be internalized to private owners and local communities. Because of its broader national importance and complex technical requirements, watershed protection should remain a state responsibility, along with other major environmental considerations. Once the state limits its responsibility to such areas as critical watersheds and nature reserves, it would be in a better position to exercise effective control. Yet, in cases where local communities can protect watershed resources more cost-effectively than the state, they should be given the opportunity and resources to do so.

Starting with insecurely held land, the state could proceed to classify forest lands into three categories: (1) land disposable to individuals, (2) land disposable to groups of individuals or communities, and (3) non-disposable land over which the state retains ownership and control. The criterion for this classification is the extent of externalities generated, both in terms of intensity and spatial distribution. Land and forests with no or only minor externalities can be conveyed to the local community to be managed communally or to individual families, depending on the local context and traditions. Finally, lands with substantial and widespread externalities over large areas should be retained under state control unless local communities at the source can be induced to take these externalities into account.

As a rule, a smaller unit of management is preferable over a larger one, unless the loss from externalities and economies of scale forgone outweigh the gains from efficient management. Certain resources and land uses, such as pastures and woodlots, may be managed with economies of scale sufficient to warrant community ownership and management, and yet not sufficient to warrant state involvement. In this context, customary property and use rights should be recognized and accommodated.

Demarcating and effectively protecting forests that play important roles in soil, water, and genetic resource conservation, and degazetting and distributing to local communities or individuals any remaining public land that the government cannot effectively protect and manage, is absolutely critical and should precede the issuance of land titles to squatters. The government could continue to be involved in production forestry in situations where it is uniquely suited to internalize externalities, or manage the resource more efficiently than the private sector. Where forests are reserved primarily for timber, but can be managed to produce non-timber goods and services as well, appropriate private or community incentives for multiple-use management may be provided as a cost-effective alternative to government ownership. It is the pervasive uncertainty and insecurity that surrounds forest concessions and land ownership, and the overhanging threat of encroachment and squatting, that discourages forest investments by the private sector and limits the success of public forest investments.

For production forests that remain in the public domain, logging rights (or rather exploitation leases) should be competitively awarded through public auctions to maximize the government's share of the rent. The leases must be sufficiently long (70 to 80 years) and comprehensive (including both timber and non-timber production) to ensure the holder's interest in forest regeneration and in protection and enhancement of non-timber goods along with timber. The lease should extend at least through the subsequent timber-felling cycle so that the leaseholders ultimately bear the consequences, and reap the benefits, of their actions.

For forests that become communal property, the transfer of rights must be secure, exclusive, and complete (including both timber and non-timber) and the communities should be free to award logging concessions or otherwise dispose of their property. It might be necessary, however, for the government to help communities strengthen their internal social organization and provide them with technical-management assistance. Community and individual investments in mixed-species plantations, woodlots, and home gardens could be actively encouraged by governments, since they would help relieve the pressure on the remaining forests by providing easily accessible fuelwood, fodder, and timber for local use. For the private sector all that is needed is credit availability, basic infrastructure, rational taxation, and incentives in proportion to the social benefits of forest investments not captured by timber and other marketed forest products.

Finally, forests that remain in the public domain, especially critical watersheds, biological reserves, and national parks, should be accorded full protection and effective enforcement of ownership by the state. This does not preclude a role for the private sector and local communities, but such a role needs to be strictly regulated and closely monitored.

CONCLUSION

The proposed solution to the problem lies in establishing appropriate property rights over the forest and all other open-access resources and in providing the rural population with better alternatives for earning a living. This implies that the solution to the forest problems is to be found outside the confines of the forestry sector. Rural development is a precondition for effective resource management, not vice versa. However, considering the political constraints to outright redistribution of assets such as land, a good place to start is with the open-access (public) resources: (1) forest lands with no significant externalities can be safely distributed and securely titled to the dispossessed, e.g., landless farmers, chronically idle laborers, shifting cultivators; (2) forest lands with localized externalities, such as local watersheds, can be made communal property provided that a community can be defined that is small and cohesive enough to manage them effectively; and (3) forest lands with regional or national externalities, such as major watershed or nature reserves, should stay under state ownership, which would be more likely to be effective over a limited area with reduced outside pressure.

Once secure property rights are established and superior alternatives for earning a living are made available, excessive deforestation and degradation could be stemmed, enabling regeneration of the forest to its optimal level. Only when workable institutional arrangements are in place will technical innovations such as intensive management, silviculture, and agroforestry have a pronounced and lasting effect on the profitability of forests.

While institutional reform of forest law and policy is a necessary condition for improved forest management and increased investment, it is not sufficient. Part of the solution to forestry problems lies outside the forestry sector and has to do with other sectoral and macroeconomic policies. The following chapter addresses these and other policy issues.

CHAPTER 12

Government Policies

GOVERNMENT TAXATION policies, public investments, and economic incentives have a critical impact, intended or not, on the forest sector. Part of the solution to the forestry problem may lie in nonforest sectoral and macroeconomic policies. For example, credit policies critically affect the level of long-term investment in tropical timber; agricultural and livestock policies that unduly subsidize forest conversion may offset forest investment incentives; and lack of successful rural development that limits alternative employment opportunities may defeat institutional reforms that seek to stem forest encroachment. High interest rates may slow or hasten deforestation depending on the capital intensity of logging, but are certain to discourage long-term investments in forest regeneration and planting. Overvalued exchange rates cause production to shift from tradeables such as timber to nontradeables such as fuelwood and subsistence goods.

One goal of this chapter is to discuss the effects of various government policies on resource depletion, forest investment, and timber supply. A second goal is to propose policy reforms that will create a favorable environment for forest investments by removing distortions and providing incentives commensurate with the forest's full social value. Three types of policies and policy reforms will be reviewed: forest sector policies, agricultural and other sectoral policies, and macroeconomic policies.

A fundamental assumption made in this chapter is that for policy reform to succeed, institutional reforms must already be in place, as discussed in chapter 11. Forests, forestlands, and contiguous agricultural lands must have exclusive and secure owners, be they individuals, communities, or governments, who are able to enforce their property rights. Enforcement problems cannot be solved overnight, following decades of open access. However, given these prior assumptions, policing costs and enforcement difficulties are expected to have

213

diminished significantly, since institutional reform minimizes enforcement costs by unambiguously defining rights and demarcating boundaries, accommodating customary rights, and increasing reliance on self-enforcement.

FOREST SECTOR POLICIES I:
GOVERNMENTS AS FOREST OWNERS

Governments play a triple role in the forestry sector, as forest owners, as regulators of economic activity, and as development agents. As forest owners, governments tend to grossly undervalue timber and to neglect non-timber goods and forest services (see chapters 4 through 6). This results in excessive deforestation and inadequate investment in regeneration and planting. Indonesia, the world's second largest exporter of tropical timber, captures only 50 percent of the rents from its logs, and 25 percent of the rents from sawn timber, leaving about 700 million dollars a year to logging companies over and above their normal profits. The implied high rate of return to investment in logging caused such a rush for concessions that by 1983, 65.4 million hectares (or 102 percent of the total production forest of Indonesia) were awarded to concessionaires (Gillis 1988). With large amounts of investment funds going into acquiring and harvesting logging concessions, and large quantities of timber coming from undervalued natural forests, it is no wonder there is little interest in investment in forest regeneration and forest plantations. In both absolute and relative terms forest investments, other than for logging, are discouraged, and there is little incentive to protect the productivity of the resource base for future harvests due to the short duration of concessions, and the perverse tax structure. As the resource base is destroyed, the production of non-timber goods and services ceases.

Degradation of the resource base caused by destructive harvesting methods, logging of marginal areas, and inadequate investment in regeneration also results in the loss of the forest's environmental services including soil, water, and genetic resource conservation. These losses may be compounded by forest fires that spread more easily because of the accumulation of combustible litter in logged-over forests. For instance, the 1983 fire in Kalimantan, Indonesia, burned down 13,000 square miles of rain forest, an area the size of Belgium. The loss of both standing timber and the growing stock was estimated at more than $5 billion (Lennertz and Panzer 1984). Preliminary assessments indicate the extinction of several species of flora and fauna, soil erosion, increased sedimentation of rivers, and microclimatic changes. Gillis attributes the severity of the fire to the logging activity of the past 15 years, concluding that:

A major social cost of extraction has not been adequately recognised and taken into account in policy decisions concerning methods and scope of timber harvests. The presence of a major, unrecognized environmental cost from logging means that the owner (the government) has priced the resource too cheaply in the past: the combination of royalties and taxes received by the owner has been well below that required to compensate for the loss in economic and social value of the forest caused by harvesting. (Gillis 1987, 7)

To enable them to act as efficient owners of the society's forest resources, governments should consider the following reforms:

1. reduction in the forest area under state ownership to manageable proportions through the institutional reforms discussed in the preceding chapter;
2. change in the procedure for awarding concessions from negotiations with the concessionaires to competitive bidding in order to maximize the government's share of the resource rent, keep logging out of marginal lands, and reduce the perceived risk of renegotiation of concession agreements (with appropriate safeguards against timber trespass and low-grading [harvesting below minimum diameter cutting limits]);
3. increase in the duration and scope of timber exploitation leases to include the value of non-timber forest products and services and encourage forest regeneration for subsequent felling cycles;
4. protection of the concession area from encroachment and enforcing the terms of the concession agreement;
5. reform of the tax system to eliminate incentives for destructive logging; e.g., elimination of regulations requiring the concessionaires or leaseholders to begin harvesting their sites by a stipulated time and a change in the tax base from "volume of timber removed" to "volume of marketable timber on site";
6. determination as to whether any harvesting of timber, fuelwood, and non-timber goods should be allowed in protective forests; specification of the areas, conditions and restrictions, and groups allowed to harvest; and development of an enforceable cost-effective system of incentives and penalties to regulate access and

use without unacceptable trade-offs between the primary "protective" and the secondary "productive" functions of these forests.
This would require research to assess the trade-offs between competing uses, predict the responses to penalties and incentives, and
evaluate the cost-effectiveness of alternative policy instruments;
7. investment in the protection, management, and enhancement of
   the state-owned forests to levels justified by social profitability;
8. introduction of an enforceable, cost-effective, and efficient system of laws and institutions to promote innovative approaches to
   the protection and management of national parks and biological
   reserves set aside for the conservation of genetic resources and the
   preservation of wilderness and recreation values.

Although an adequate economic return is not the only justification
for such conservation areas, such a return can often be generated to
defray protection and management costs by charging access fees to
tourists, scientists, pharmaceutical and food companies, downstream
farmers, irrigation systems, and others who benefit from such conservation areas. For example, a World Bank-assisted irrigated rice project
in Sulawesi, Indonesia, funds the protection and management of the
3,200-square-kilometer Dumoga National Park, which covers the watershed catchment area for the Dumoga irrigation system. The irrigation project benefits from reduced maintenance costs and increased
dry-season water availability that result from the park's management.
The park preserves and protects the forest's genetic diversity and recreational value (World Bank 1987).

FOREST SECTOR POLICIES II:
GOVERNMENTS AS REGULATORS OF ECONOMIC ACTIVITY

As regulators of economic activity, governments may control the use of
forests and forestland by the private sector and by local communities in
order to promote social objectives such as equity, stability, and national
security; to mitigate market failures such as externalities, shortsighted
behavior, irreversibility, and market imperfections; or to provide public
goods that would not be provided by the free market.

The forest sector has been a target of government regulation and
incentives for growth. Forests represent a classic case of pervasive externalities or spillover effects. Tropical forests conserve soil, water, and
genetic diversity. They have aesthetic, scientific, and recreational value.
Forest degradation may result in soil erosion, flooding, sedimentation

of water bodies, loss of species, scarred landscapes, and unfavorable microclimatic changes such as a rise in local temperature, increased frequency of droughts, and heavier, more destructive rainfall. These are all external effects that an unregulated market will largely ignore or grossly undervalue, creating a demand for government intervention to internalize them through taxation, regulation, or state ownership.

Tropical forests take decades to grow; a market oriented toward short-term returns will lead to overexploitation of existing forests and underinvestment in forest regeneration and planting. Such myopia is especially likely under conditions of insecure tenure and political or economic uncertainty. Even under conditions of certainty, a free market may underinvest in the preservation of the resource base for future generations; therefore, there is a role for the government as a guardian of the resources for future generations. Moreover, even if it is economically sound under current price-cost expectations, the destruction of tropical forest might still be socially suboptimal because the extinction of species and land degradation might result in permanent and irrevocable foreclosure of options.

Markets in general, particularly in developing countries, are far from complete or competitive, and are often riddled with local monopolies and monopsonies, preemptive marketing arrangements, and other uncompetitive practices. Imperfections in the capital market severely affect the forest sector. Because of their long gestation period and the consequent timelag between expenditures and returns, forest investments depend on investors being able to borrow at the prevailing interest rate against future harvests, 50 to 70 years hence. Highly fragmented and distorted by interest-rate ceilings and credit rationing, capital markets in developing countries actually discriminate against rural areas and long-term investments.

Finally, many of the forest's environmental services, such as watershed protection, genetic diversity, climatic effects, and recreation, are public goods that cannot be profitably provided by a free market. Exclusion is either not possible or socially suboptimal as, for example, in "free rider" situations in which more people sharing in the good raises social benefits without increasing social costs. Therefore, public goods such as watershed protection, national parks, and biological reserves can best be provided by the state and financed through general taxation.

Thus, there is considerable scope and need for a coherent government forest policy that transcends the government's role as forest owner. Such a policy should aim to (1) internalize the externalities generated by private forest activities such as timber harvesting and private forest investment; (2) ensure the availability of financial resources to carry out productive forest investments by the private sector and local communities; and (3) finance and supply public forest goods and services that cannot or should not be supplied by the private sector. Because they involve the internalization of externalities, the institutional reforms proposed in chapter 11 would reduce but not eliminate the need for greater government regulation. Moreover, the security of tenure and reduction in uncertainty resulting from institutional reforms would increase access to institutional credit and the availability of private funds for investment.

Even with institutional reforms, however, private forest investments will continue to be at socially suboptimal levels because (1) private financial rates of return do not reflect all the social benefits that forest investments generate, and (2) their characteristically uneven cash flow, large size, and long gestation periods make these investments highly risky. For these reasons, governments throughout the tropics have taken over control of forest resources. In theory, governments can manage forests for multiple uses and can invest in forests at optimal levels because they have broader social objectives, longer planning horizons, lower discount rates, and a greater ability to pool risks. In practice, however, government ownership and public investment are inefficient at internalizing externalities and attaining socially optimal levels of investments.

With the exception of public goods such as national parks and nature reserves, decentralized ownership with an appropriate incentive structure and financing mechanisms is usually a more efficient means to attain the socially optimal level of forest investments and encourage multiple-use management. Where private forest investments generate public benefits, the government should link commensurate incentives such as tax exemptions and subsidies to these benefits to encourage forest investments to rise to a level consistent with their long-term economic and social profitability. For instance, the tax structure should favor natural forest management over plantations, mixed-species plantations over single-species plantations, and single-species plantations over erosive crops such as corn and cassava (see Panayotou 1991 for further details on land-use and slope-based taxes). Plantations should be taxed or promoted in proportion to their net social and environmental impact on the water table, soil erosion, nutrient depletion, etc. Logging companies could be provided with incentives to set aside part of their concessions as nature reserves (for conservation purposes) and extractive reserves (for social purposes) and manage the rest on a sustainable basis.

With regard to the financing of private forest investment, governments have a wide range of instruments including cofinancing, establishment of guaranteed funds to reduce risk, sectoral and global loan programs, and insurance against pest outbreaks and forest fires. According to the InterAmerican Development Bank (IDB), governments should consider actions to:

1. develop new and innovative ways to adapt national financing mechanisms to the circumstances of the sector without ignoring basic financial, management, and economic principles; and
2. explore with international lending institutions means to improve the effectiveness of sector investment programs using international loan and technical assistance funding, possibly including sector and global loans, cofinancing mechanisms, loan guarantees, and direct technical assistance for project identification, preparation, and institutional strengthening. (McGaughey and Gregersen 1983, 27)

High interest rates are often blamed for the low levels of private forest investments (Leslie 1987), but according to the IDB, "the most critical loan conditions for forestry projects are the grace and disbursement periods and the foreign exchange content. As long as financing is available, the rate of interest, except at very high levels, is less critical to project success" (McGaughey and Gregersen 1983). Proposed policy incentives include the doubling of grace and disbursement periods, the use of repeated global loans, increasing the foreign exchange component of project loans to cover indirect costs and to compensate for the initial cash-flow deficiencies, and shifting cultivation toward tree species with shorter financial rotations. However, some of these proposed policy responses, such as the move toward faster-growing tree species, may be neither necessary nor desirable if non-timber forest products and services are integrated into both natural forest management and plantation investments through appropriate institutional arrangements and economic incentives.

FOREST SECTOR POLICIES III:
GOVERNMENTS AS DEVELOPMENT AGENTS

As development agents, governments attempt to stimulate economic growth by providing infrastructure, investment incentives, protection of infant industries, extension of new technologies, and even strategic public investments in selected industries that could serve as leading growth sectors. In most developing countries, governments play a more active role in promoting economic development than is the case in developed countries.

Governments find ample reasons for treating forestry as a development sector. First, in many tropical developing countries in early stages of development, forests are the major, if not the only, source of investible surplus and foreign exchange for the import of capital equipment necessary for development. Second, tropical forests also serve as a source of land for rural, agricultural, and livestock development. Finally, tropical forests are a source of readily available raw materials, particularly timber, on which to base local industries. A related development objective often realized through the forest sector is the generation of local employment and value added from primary resource exports by promoting more local processing of forest products.

The "vent for surplus" attitude toward tropical forests encourages governments to promote the liquidation of forests and the channeling of revenues into other sectors that are considered to be more modern or to have better growth potential. While this is justifiable up to a point, in their haste to maximize development returns governments often overlook the opportunity to manage tropical forests on a sustainable basis. They fail to recognize that forests can yield a perpetual stream of benefits from non-timber goods and services as well as timber, and they ignore the cost of compensating for lost services. The result is excessive deforestation and underinvestment in protection, management, and regeneration of natural forests.

This outcome is reinforced by the perception of natural forests as an inexhaustible reservoir of land for expansion of agricultural and human settlements. Again, forest conversion is justified up to a point, beyond which the social costs outweigh the benefits. As agriculture expands from the lowlands toward more marginal and fragile uplands, cultivation becomes increasingly less productive and less sustainable, while the protective functions of the tropical forest assume increasing importance for both the uplands and the lowlands. To determine the socially optimal land use, the net present value of alternative uses must be compared. The greater value of a perpetual stream of timber and non-timber goods and environmental services from forested land should be obvious, compared with the benefits of unsustainable agriculture or cattle ranching made possible only by irreversible forest loss. Yet governments have promoted resettlement schemes, cattle ranching, and other forms of forest conversion, all in the name of development. Examples range from the heavy subsidization of cattle ranching in Brazil to transmigration projects in Indonesia; from coffee plantations in the Ivory Coast to cassava production in Thailand. In the Brazilian Amazon, where livestock projects were promoted through tax holidays, tax credits, offsetting tax losses, and subsidized credit, the net present value of government-assisted cattle ranches was found to be negative (Repetto 1986), and indeed the low net returns from many of these policies caused the Brazilian government to suspend many of the tax advantages in 1988.

Full valuation of timber and non-timber goods and services in a multiple-use management framework might have prevented or checked the "development" policies that permitted or encouraged such massive conversions of tropical forests to unsustainable or marginal uses and the consequent misallocation of other scarce resources, including capital, labor, and government revenues.

The promotion of local processing of primary exports such as timber is also a well-meant and often justified policy. Naturally, governments react to the large discrepancy between the export price they receive for their unprocessed timber and the higher price that plywood, veneer, and other wood products derived from their timber receive in the world market. It is an understandable reaction to want to ban exports of unprocessed timber and to promote domestic processing into plywood and other wood products. It is also an understandable reaction to raise the protection of and subsidies to the local plywood mills when developed countries impose import tariffs on plywood and other processed wood products from developing countries. In spite of the fact that these reactions have good intentions and are understandable, they do not always have the expected favorable outcomes.

In many cases, plywood mills established in response to these industrialization incentives are inefficient. In Ghana, for example, plywood mills require 22 percent more timber per unit of output than Japanese mills, a ratio that translates to 22 percent more logging for the same level of output. In Indonesia, according to Repetto (1986, 72), "the government sacrificed $20 in export taxes in order to lose $11 in value added on every cubic meter of logs sawn domestically." The point is not that forest-based industrialization should not be promoted, but that incentives must be commensurate with the expected net social benefits not captured in the private benefit-cost calculus. The gain in employment and local income should be weighed against the loss of export taxes, the waste of timber, and the opening up of more forest areas to encroachment, with the consequent loss of non-timber goods and services and future timber. Consideration of non-timber goods and services may well serve to moderate the perceived need for local employment creation through forced forest-based industrialization.

SECTORAL AND MACROECONOMIC POLICIES

We have already seen that certain agricultural and industrial policies may induce excessive conversion of tropical forest lands into other uses that are not always more productive or sustainable. This proposition can be generalized by stating that the relative profitability of forest investments (including the passive investment of keeping forest lands as forest) depends not only on forest policies and forest investment incentives but also, and critically, on non-forest policies and incentives.

The returns to non-forest investment (i.e., in agriculture, industry, or services) reflect the opportunity costs of forest investments. Government policies that artificially raise the returns to the former are equivalent to policies that lower the returns to the latter. For example, agricultural price supports, agricultural input subsidies, supply of irrigation water free of charge, industrial investment promotion, and import substitution through tariff protection combine to bias the allocation of limited investment funds against forest investments. The solution is not a sweeping elimination of these policies or introduction of offsetting subsidies for forest investment, but rather the realignment of these policies to correspond with each sector's or activity's long-term economic and social profitability.

Finally, macroeconomic policies not aimed at a particular sector but applying equally to all sectors may have different effects on different sectors because of the disparate composition, structure, and planning horizon of these sectors. For example, overvaluation of the exchange rate would affect the more export-oriented sectors more seriously, which in some tropical countries includes the forest sectors. Within the forest sector, overvalued exchange rates cause a shift of production and investment from tradeables (e.g., timber, rattan, Brazil nuts) to nontradeables (e.g., fuelwood, locally consumed non-timber goods, and subsistence production). Whether this is beneficial for countries experiencing excessive rates of deforestation depends on the net effect of the reduction of logging and damage from logging, on the one hand, and increase in fuelwood collection and slash-and-burn cultivation on the other.

Other macroeconomic policies that may profoundly affect forest investments are monetary policies that lead to a change in interest rates and credit policies that impose interest-rate ceilings. High interest rates generally increase logging and discourage forest investments relative to other investments, which tend to have shorter gestation periods. Credit policies that impose interest-rate ceilings result in discrimination against forest investments, which are longer-term and bear greater risk.

Minimum wage policies for industry reduce industrial employment and increase rural underemployment, a situation that can exacerbate forest encroachment. Industrial incentives in favor of urban areas that discourage rural industry have similar effects (Panayotou 1990). General development policies that provide alternative employment opportunities to forest squatters and shifting cultivators, on the other hand, may ameliorate these negative conditions.

CONCLUSION

The effects of various government policies on resource depletion, forest investment, and timber supply have been overlooked. Governments have a unique fiduciary role to play in the setting of incentives to encourage long-term sustainable production of forest resources; to a great extent this role has been subordinated by other priorities.

As forest owners, governments have undervalued timber and neglected non-timber goods and services. Potential policy reforms include reduction in the forest area under state ownership, change in the structure and awarding of concessions, reform of the tax system, protection of and investment in appropriate forest areas, and creation of incentives for valuation of forest services. As regulators of economic activity, governments would be well advised to reform existing investment codes within the forest sector as well as to promote multilateral and bilateral investment in sustainable forest production.

As development agents, governments have tended to emphasize short-term development gains to the detriment of long-term resource sustainability, hence the emphasis on timber production and forestland conversion to other uses. A more enlightened view of economic development could focus on maximizing long-term benefits from tropical forests through multiple-use management that pays due attention to non-timber products and services in addition to timber.

Finally, as authors of macroeconomic policy affecting the forest sector and the general economy, governments must seek to evaluate the effects of exchange-rate policy, fiscal spending, and interest rates on short- and long-term productivity of tropical forests. Policies determined at the level of Ministry of Finance, Central Bank, or Executive Office can enhance or detract from the viability of the forest resource.

CHAPTER 13

International Cooperation

WHILE INSTITUTIONAL and policy reforms by individual governments would make a major contribution to rationalizing the exploitation of forest resources and improving the economic environment for forest investments, these reforms may be partially frustrated by the continuing supply of undervalued timber from other countries that mine rather than manage their natural forests. Countries that import tropical timber stand to benefit as much as exporters from the sustainability of timber supplies that would result from full valuation of all forest goods and services. Importers would benefit directly as well from the conservation of genetic resources and improved economic and natural environments. These benefits would be larger and more likely to materialize if there were coordinated action by the producing and consuming nations, as a group, to rationalize the use of tropical forests on which the sustainability of the trade in all forest products ultimately depends.

There is scope and need for international cooperation in four respects:

1. to coordinate action so that institutional and policy reforms in one country are not frustrated by fears of losing market shares to other countries;
2. to ensure adequate financial flows and technical assistance for forest investments;
3. to help finance the conservation of genetic resources and other international public goods that generate benefits beyond the borders of the producing nations; and
4. to fund and coordinate mutually beneficial research and development programs and to share information, research findings, and

experience concerning natural forest management, forestry investments, and institutional and policy reforms.

POLICY COORDINATION

In principle, extraction of more rent from timber harvesting should not affect timber production and prices as long as it is done through competitive bidding or through a profit or stumpage tax. In reality, however, a modest short-term increase in timber prices may result because of the consequent withdrawal from marginal areas that are now kept in production because of uncollected rents. This price effect would be a modest one because these areas are not highly productive in terms of timber, although they may be very productive in terms of non-timber goods and environmental services. In the medium and long run, increased security of tenure, reduced uncertainty, full valuation of timber and non-timber goods and services, and their internalization through economic incentives should result in significantly increased investment in natural forest management and plantations. The effects of such investment would include moderation of timber prices.

However, individual countries could voice justifiable concern that if they make unilateral reforms, such as increased taxation of timber rents or requirements that concessionaires set aside conservation areas, the logging companies would be likely to shift their operations to other countries with lower tax rates and less regulation, or raise their prices with consequent loss of market share to other countries. This concern would be particularly strong in countries with a struggling domestic timber-processing industry. To prevent such short-term localized concerns from forestalling policy changes that are critical to the long-term sustainability of the timber trade, an international agreement (convention) and cooperation are necessary to coordinate action and cushion the short-term impact.

FINANCING FOREST INVESTMENTS

Investment in forest management is critical in this process. As prior chapters have demonstrated, a scarcity of financial resources for forest investments results from (1) the undervaluation of timber, neglect of non-timber goods and services not accruing to private investors, and insecurity of ownership; (2) the special characteristics of forest investment such as long gestation periods, uneven cash flows, large size of investments, and high risk and uncertainty; (3) the lending conditions of international lending agencies, particularly the grace, repayment, and disbursement periods and foreign exchange components, are ill suited to forest investments; and (4) the lack of well-prepared productive forestry investment projects. According to the InterAmerican Development Bank:

> [I]nternational and regional sources have assisted the [forest] sector only marginally as a share of their total lending . . . because of a shortage of their total lending . . . [and because] forest development often has been narrowly conceived as commercial forestry and forest industry. More broadly conceived forestry includes social and conservation forestry which contributes significantly to the well-being of rural inhabitants. (McGaughey and Gregersen 1983, 29)

International cooperation could help address some of these problems by exploring alternative financing instruments for the forestry sector that include cofinancing schemes (among international and regional lending institutions, export finance agencies, and commercial banks), global loan programs, and adjustments of repayment, grace, and disbursement periods to better fit the cash-flow profile of forestry investment projects. International cooperation can also play an important role in technical assistance for identification, preparation, and implementation of viable forest investment projects.

INVESTING IN THE GLOBAL COMMONS

Finally, while many of the environmental services of tropical forests, such as soil and water conservation, are critical for the sustainable development of tropical countries themselves and should be financed accordingly, other services of tropical forests are truly international public goods. Preservation of wilderness, biological diversity, and global ecological balance all generate benefits beyond the borders of the countries that produce these services. In fact, the producers of these services—primarily the tropical timber nations—would benefit less than others because these services are basically "luxury" goods (i.e., in high demand in rich countries and low demand in poor countries). The highest benefits would accrue to the developed countries of the European Community, the United States, and Japan, the major importers of tropical timber. These countries should be prepared to help fund the provision of these services through the creation of national parks and biological reserves in the producer nations. Preservation of unique genetic resources and rare ecosystems would also help to reduce the opposition to timber trade on environmental grounds and ensure its long-term economic and social viability. In this context, the feasibility of a Tropical Forest Conservation Bank with major contributions from timber-producing and consuming nations should be explored. Such a bank could compensate tropical countries for forgoing forest conversion (and timber production) to preserve their forests as biological reserves and producers of global environmental services.

PROMOTING CROSS-BORDER RESEARCH

Despite the many differences between and within the three continents, two seemingly contradictory observations are striking: (1) the commonality of problems facing rain forests throughout the tropics, and (2) the relatively limited transfer of knowledge and experience with management systems, institutional arrangements, and government policies across countries. International cooperation can play a critical role in identifying common problems, transferring knowledge and experience, and designing research projects that generate transferable knowledge without sacrificing important regional or local dimensions. Economies of scale and scope can be gained from pooling research resources (financial, scientific, and experimental) to address problems that cut across several countries or can give rise to generalizable results.

CHAPTER 14

An Agenda for the Future

THE WORLD'S tropical forests are fast disappearing. Studies conducted around the globe in countries such as the Ivory Coast, Costa Rica, Thailand, and India all report massive deforestation. While conservative estimates put deforestation in the tropics range at 11 million hectares per year, other reports place the figure much higher—at 19 million hectares annually. In either case, there is ample cause for concern, particularly when considering the numerous goods and services that tropical forests have to offer.

We recognize that the crisis of deforestation is an almost universal accompaniment of the early, least-coordinated period of development and was experienced by the now developed countries in their time. When traditional, predominantly noncash societies, who customarily lightly harvest a wealth of goods from tropical forests, enter the wage economy, the forests' value plummets, as traditional uses are abandoned for perceived superior alternative products in the market. Simultaneously, overexploitation of marketable forest products, such as rattan and bushmeat as well as timber, is fostered by the absence or contravention of traditional constraints, introduction of more exploitative technologies such as the chainsaw and shotgun, and insecurity of tenure brought on by disregard for traditional ownership rules. Successful development eventually provides alternative employment for forest edge dwellers, often far from the forest, and renewed appreciation of forests as a more affluent generation comes to require their aesthetic, traditional, recreational, and existence values. The challenge therefore is to establish the true value of forests during the crucial early stages of economic development.

The reasons for deforestation are varied and complex, but are rooted in a linked series of market and policy failures that lead to overvaluation of the timber industry, undervaluation of the non-timber goods and services that forests provide, and significant inefficiencies in the allocation of forest resources. As the owners of more than 80 percent of the world's tropical forests, the governments of many developing countries have been unable or unwilling to manage the tropical forests on a sustainable basis. Rarely do policy makers consider managing their country's tropical forests for multiple uses. Instead, the current debate over the use of tropical forests pits economic growth directly against conservation; exploiting the forests for timber is often considered the growth-oriented option, while production of non-timber goods and services is seen as environmentally sound but unprofitable. However, multiple-use management for timber and non-timber goods and services can both maximize economic growth and conserve the forest's value for the future.

Multiple-use management enables policy makers to maximize present and future benefits according to the type of forest resource under consideration. For example, an undisturbed moist forest could well yield the highest net present value through total protection; the combined discounted value of genetic resources, non-timber goods, services (including watershed protection, CO_2 sequestration, etc.), and future timber yields from mature hardwoods could mean that the optimal management scheme is total conservation, subject to periodic re-evaluation of the costs and benefits involved (see chapter 7). By the same token, a forest consisting of parcels (effectively stands) of differing age following initial timber harvesting might be managed to yield maximum value through encouraging both timber and non-timber species, to be produced, in succession or simultaneously, over a 35- to 50-year rotation; the parcels might be selectively harvested and manipulated by liberation thinning to favor those tall-growing pioneer and light hardwood climax species that comprised the building phase in the original virgin forest. In this second case, biodiversity and environmental services do not exhibit long-term values sufficient to justify full protection; instead, the production value of a variety of forest goods and services, including some timber, is maximized over time, and investment in the resource is planned accordingly.

In order to accomplish these or other types of optimization, forestry policy must reflect the link between scarcity and prices. At present, not only are most forest products and services not priced, but even timber that is an internationally tradeable commodity is priced below its true scarcity value due to implicit and explicit subsidies and institutional failures. Uncollected resource rents, subsidized logging on marginal and fragile forestlands, and volume-based taxes on timber removal encourage high grading and destructive logging. Government-issued forest concessions are typically too short to provide incentives for conservation and replanting. Failure to value non-timber goods and services results in excessive deforestation, conflicts with local communities, loss of economic value, and environmental damage. Artificial support of local processing of timber often leads to inefficient plywood mills, excess capacity, waste of valuable tropical timber, and loss of government revenues. Replanting subsidies often end up subsidizing the conversion of a valuable natural forest to inferior mono-species plantations, with the associated loss of the value of both tropical hardwoods and biological diversity.

A Policy Agenda

What is needed is a reform of current forest policies to encourage efficient harvesting and processing and to promote investments in forest regeneration and conservation. A forest policy reform might include the following elements, most of which can be implemented internally by individual countries:

1. Reclassify forestlands into (a) land disposable to individuals: forestlands with no significant externalities can be safely distributed and securely titled to the dispossessed (e.g., landless farmers, chronically unemployed laborers, shifting cultivators); (b) land disposable to groups of individuals or communities: forestlands with localized externalities, such as local watersheds, can be made communal property provided that a community small and cohesive enough to manage them effectively can be defined; and (c) nondisposable land over which the state retains ownership and control: forestlands with regional or national externalities such as major watersheds or nature reserves should stay under state ownership, which would be more likely to be effective over a limited area with reduced outside pressure.

2. Change the procedure for awarding concessions from negotiations with the concessionaires and licensing with nominal fees to competitive bidding, in order to maximize the government's share of the resource rents, to keep logging out of marginal lands, and to reduce the perceived risk of renegotiation of concession agreements. Concessionaires should be provided with financial instruments for accumulating equity through forest investments that are transferable and marketable to encourage them to invest in conservation and reforestation.

3. Increase the duration and scope of the exploitation leases suffi-

ciently to internalize non-timber forest products and services and to encourage forest regeneration for subsequent felling cycles.

4. Protect the concession area from encroachment and enforce the terms of the concession agreement.

5. Reform the taxation system to eliminate incentives for destructive logging.

6. Determine whether any harvesting of timber, fuelwood, and non-timber goods should be allowed in protection forests, and if so, specify the areas, set the conditions and restrictions, define who should be allowed to harvest, and devise an enforceable, cost-effective system of incentives and penalties that would regulate access and use without unacceptable trade-offs between the primary "protective" function and the secondary "productive" function. This would require research and experimentation in assessing trade-offs between competing uses, predicting behavior in response to penalties and incentives, and evaluating the cost-effectiveness of alternative policy instruments.

7. Invest in the protection, management, and enhancement of the state-owned production forests based on strict criteria of social profitability.

8. Devise an enforceable, cost-effective, and efficient system of laws and institutions to stimulate innovative approaches to the protection and management of national parks and biological reserves set aside for the conservation of genetic resources, the preservation of wilderness, and recreation values.

9. Promote private forest investments through an appropriate incentive structure and financial mechanisms, such as cofinancing of long-term loans; longer grace, disbursement, and repayment periods; establishment of guarantee funds to reduce risk; and insurance against pest outbreaks and forest fires.

10. With regard to public benefits generated by private forest investments, such as downstream irrigation benefits, provide commensurate incentives—tax exemptions and subsidies linked to these benefits to bring forest investments to a level consistent with long-term economic and social profitability. For instance, the tax structure should favor natural forest management over plantations, mixed-species plantations over single-species plantations, and single-species plantations over erosive crops such as corn and cassava. Eucalyptus and pine plantations should be taxed or promoted in proportion to their net social and environmental impact on water table, soil erosion, nutrient depletion, etc. Logging companies could be provided with incentives to set aside part of their concessions as nature reserves (for conservation purposes) and extractive reserves (for social purposes) and to manage the rest on a sustainable basis.

11. Recognize, accommodate, and, where necessary, regulate the customary rights of access to and use of forests by local communities; their physical presence in the forest and their intimate knowledge of the local ecology can be of immense value in the protection and regeneration of the forest and the harvesting and use of non-timber products. Such recognition must take account of the dangers of over-exploitation of the resource by the same people either through commercial sale or as a result of population growth. In this regard, the availability of alternative employment and income opportunities is critical for sustainable resource use.

These reforms should be strongly supported by both the commercial forestry sector (both producers and consumers) and developing country governments because they will ensure sustainable supplies of tropical hardwoods and will transform tropical commercial forestry from an extractive industry into a sustainable economic activity with considerable private and social net benefits. While higher hardwood prices may be perceived as running against the short-term interests of commercial forestry (especially by the importers and consumers), the long-term benefits appropriately discounted exceed any short-term costs. Unless higher prices are paid for tropical hardwoods, there can be no conservation, and without conservation, there can be no sustainable supplies. The waste, inefficiency, and damage to regeneration are currently so great that, by instituting these reforms, almost every party involved would be made better off.

A RESEARCH AGENDA

Certainly, some policy reform can be implemented immediately on the basis of existing data. The research for this book has provided convincing evidence that models for sustainable production of tropical hardwoods, differing according to regions and countries, do exist; that the economic prospects for hardwoods and other products of tropical evergreen forests are promising, provided that certain critical institutional policy changes are made, and that the non-timber goods and services provided by tropical forests are an integral component of viable economic and management models.

Considerable further research is required, however, to fill continuing basic knowledge gaps. The most critical gap in knowledge is the absence of standardized, unfragmented data relative to the conditions necessary and sufficient to sustain production of various forest goods and services. Major gaps include the following:

1. Standardized data for global comparisons of costs and benefits of

different economic, social, management, and institutional options do not exist.

2. The interdependence among different elements—biological, ecological, social, economic, and institutional, which together must determine the parameters for securing sustainable production—cannot be reliably determined from existing data. In particular, such data are required from representative forest areas in each of the major regions of the tropics, including Africa, Latin America, and Asia/Pacific.

3. Data, standardized by area and time, on production and harvesting rates, existing trade patterns, employment levels, and values for the total of non-timber goods from representative forest areas hardly exist for any region.

4. Comprehensive data of the same kind do not exist for any single non-timber good of the tropical forests, with the possible partial exception of rattan.

5. Reliable field data for realistic calculation of the value of services provided by forests—notably conservation of water, soil, and biological diversity—do not exist for representative forest areas.

6. In several instances, methodologies for calculating the value of forest services have not yet been developed. Notable is the lack of satisfactory and operational methodology for valuation of biological diversity, and of climatic influences.

7. Without the foregoing, there is no possibility to devise management protocols either to optimize sustainable production of goods and services from a forest or to optimize sustainable harvesting rates.

In Asia, steps are being taken to address these issues. A discussion among resource scientists from the region, held in Bangkok in 1989 (Anon. 1989), has laid the groundwork for research toward these ends at representative forest sites throughout tropical Asia, using methodologies that would permit statistical comparison. Their principal objectives are (1) to estimate the total value of the chosen forests through field investigation and to estimate optimal sustainable yields and (2) to develop and test models for optimizing yields of the most-valued goods thereby identified. Research has already begun at several sites.

Specific data gaps range over all aspects of forest production, as well as the socioeconomic factors that influence the establishment of management systems to promote sustainable production.

Silvicultural Data and Analysis

Production Data. A basis for standardized comparisons of silvicultural options worldwide is provided by recognition that the diversity of tropical forest species can be classified into four broad categories according to growth and reproductive characteristics, and the fact that producers of timber and non-timber goods, including vines, all belong to these same categories (see chapter 3). Immediately needed are data on growth rates and wood-volume increments within each category, relative to the range of soils and rainfall seasonality occurring in the humid tropics. These data will allow worldwide standard comparisons of past silvicultural successes and failures and, hence, prescriptions for the most promising options for each region. Knowledge of the eco-physiological determinants of optimal tree and non-timber species performance remains patchy, and research to this end should be strengthened.

Interdependencies. The interdependencies of timber and non-timber-producing forest species have yet to be examined sufficiently. Needed most are data on the degree of ecological competition between the producers of wood and of other forest products, and the extent and means by which current methods of timber exploitation and sil-vicultural management influence such competition.

Silvicultural experiments should be initiated in which species yielding non-timber products, including species providing services such as food for game, are accorded the same silvicultural treatment as the timber producers, to evaluate the impact of such treatment on overall forest production, profitability, and demand for forest goods and services.

Biological interdependencies can only be analyzed by thorough de-mographic research. For trees, this requires establishment of sample plots large enough to include samples sufficient for demographic anal-ysis of the majority of species. Fifty hectares have been found satisfac-tory in species-rich rain forest (Manokaran et al. 1990). Such samples additionally serve as controls for silvicultural experiments, as refer-ences and databases for animal demographic, socioeconomic, and pro-duction studies, and for phonological and other silvicultural research.

Enrichment Planting. The limited success of enrichment planting of tropical hardwood forest may be attributed to management and super-vision problems arising largely from inadequate or inappropriate in-centives. The socioeconomic and institutional reasons for these shortcomings should be defined and addressed, and further experi-ments in enrichment planting should be attempted. Priorities for ex-perimental enrichment planting should include combined plantings of building-phase light industrial hardwoods (as shelter for interplanted quality hardwood species) and interplanting of timber-producing spe-cies with non-timber producers, including rattan and specialist crops.

Plantations. Stringent ecological comparisons of single- and multiple-species plantations under similar site conditions are required. Initially, commercial plantations and home gardens can provide a basis for comparison, but experiments should be initiated in which the same array of species are grown in monoculture, and according to various mixed-plantation designs. Mixed-species plantations entirely of timber producers, and mixtures with non-timber producers, are required. Evaluation should include consideration of productivity for various goods, and its predictability; effects on soil and water; and the degree to which local and regional commercial and socioeconomic expectations are met.

Socioeconomic and Institutional Data and Their Analysis

Areas considered representative of permanent forests should be identified in one or more nations in each region of the humid tropics. In each, standardized data should be gathered with the aim of:

1. establishing estimates of the net present value of forests of different kinds (including human-made forests), incorporating the totality of the goods and services they render;
2. identifying socioeconomic interdependencies of forests, on local as well as regional and international levels;
3. devising means to define and evaluate externalities such as conservation of genetic resources, protection of watersheds, climatic effects, and carbon sequestration;
4. describing the social, economic, political, and institutional impediments to sustainable management of hardwood forests within the broader context of the added value of forests and the role of forests in regional development; and
5. defining the institutional requirements for sustainable hardwood forestry.

From these analyses, a series of regional studies of integrated forest policy could be produced, which in turn would serve as a basis for a global model for forest policy development. Using these research resources, prescriptions could be made for individual regions and countries.

Valuation. Net present values are required for non-timber products and services at realistic potential production levels, as well as for timber production. Analysis of the commercial potential of promising non-timber goods must be undertaken for each region, and incorporated with data on productivity and cultural requirements. Net present values of managed natural forests and plantations should be compared under differing sites and sociopolitical conditions. Silvicultural data on growth and yield of different categories of useful forest plants, under different site conditions, is of highest priority. Valuation of the totality of goods and services yielded by forests under different forms of silvicultural management, including plantations, is required. Particular attention should be paid to economic gains from harvesting timber in mixture with species that produce non-timber goods between felling cycles.

Forest goods that do not enter the cash economy, and those traded locally and nationally without being taxed or otherwise assessed and entered into statistics, require valuation. Standard means, direct or indirect, must be devised to value services, including water conservation and quality, soil protection, conservation of biological (genetic) resources, and climatic influences.

The valuation of timber and non-timber production should net out all relevant costs. With respect to timber production, these include costs of damage to residual stock, costs of soil deterioration, and net nutrient loss after harvesting. On the non-timber side, such valuation should include estimation of the opportunity costs of harvesting particular species (e.g., the removal of nutrients and seeds and its effect on surrounding plant demography), as well as investment requirements for transportation and marketing systems. Impacts of exploitation methods on non-timber goods and services, particularly those of socioeconomic importance to the rural economy, must be analyzed with the aim of minimizing and, wherever possible, eliminating adverse effects.

Social Values. To ensure full valuation of the goods and services provided by the forest, analyses should include the following:

1. total employment that is provided directly or indirectly by the forest, including the contribution of timber and non-timber sectors;
2. patterns of trade in all forest goods, and distribution of profits with a view to identification of inefficiencies and inequitable distribution, particularly where they may affect incentives for conservation and sustainable management of the forest resource;
3. social implications of different forest management options, including different silvicultural prescriptions, and single and

multiple-species plantations—particular priority should be given to the impact that various management options may have on the economy and employment opportunities of those communities that influence the long-term security of the forest resource;

4. property rights, with a view to identifying those that may provide maximum incentives to manage the forest resource on a sustainable basis;
5. externalities in the different sectors of the forest economy in a search for means to internalize them through institutional changes;
6. unsatisfactory institutional arrangements, including laws that require forest removal as a condition of land tenure, even if alternative uses are unsustainable, or economic conditions and institutional arrangements that encourage conversion of indigenous forest to plantation even when the latter is less productive.

The case for government ownership would appear to be incontrovertible in the case of forests such as major watersheds or genetic diversity preserves, where the benefits cannot be internalized at less than a national scale. Options for international sharing of responsibility and costs, as a means to internalize externalities at national levels, should be explored (Panayotou 1992).

A First Step in Policy and Research

A first step to gathering this kind of information is already underway. "Long-term management research sites" (LTMRS) are operative in tropical Asia and Central America to generate data for an interdisciplinary analysis of biological, anthropological, and economic factors governing optimization of forest goods and services and conservation of biodiversity. Sites in Thailand, Malaysia, Sri Lanka, and Panama have been identified and constitute the initial locations for long-term research; other sites are in the pipeline.

Research goals at these and future sites include the following:

1. recording the full range of current uses of tropical forests, including patterns of use and marketing of goods;
2. documenting aspects of biodiversity;
3. estimating the likely future value of existing resources;
4. estimating the quantity and rate of production for those species

important economically or ecologically to biodiversity in the
forest;

5. examining positive and negative interdependencies among species, with a focus on the dynamics of biodiversity maintenance
and species productivity;

6. developing and testing predictive models for optimizing sustainable production of the goods and services deemed most significant on a case-by-case basis.

This effort has been institutionalized recently with the establishment
of the Center for Tropical Forest Science (CTFS) under the Smithsonian Tropical Research Institute. The aim of CTFS is to facilitate
international applied social and natural science research to advance the
sustainable use and management of tropical forests and the conservation of their biodiversity.

CONCLUSION

Multiple-use management of tropical forests is but one potential solution to the problem of disappearing forest resources. As such, it deserves more intense recognition, research, and where immediately
beneficial, implementation. It is not by timber alone that forests can
generate substantial returns, yield social benefits, and enhance development in those countries that harbor them.

But mixed-use management, in isolation, cannot halt the accelerating disappearance of tropical forests worldwide. As international concern increases for the global forest resource, policy makers, research
scientists, and local communities must all cooperate in deepening the
policy agenda and expanding the knowledge base in order to respond
to this concern constructively. Deepening the policy agenda means not
only that individual governments should eliminate those policies that
clearly create inefficiencies but also that governments of developed as
well as developing countries must maintain a policy dialogue in order
to explore new options for optimizing a variety of returns from tropical
forests. Expanding the knowledge base includes setting aside more
research sites in tropical forests for examination of management
options and exploration of production priorities. As the benefits flow-
ing from forests are both local and global in scale, so must the costs of
sustaining them be borne both locally and globally. In the final analysis,
the benefits from sustaining tropical forests for both the current and
future generations far surpass the costs.

Bibliography

CHAPTER ONE

Grainger, A. 1987. "The Future Environment for Forest Management in Latin America." In *Management of the Forests of Tropical America: Prospects and Technologies*. Institute of Tropical Forestry/USDA Forest Service, Washington, DC.

Leighton, N., and N. Wirawan. 1986. "Catastrophic Drought and Fire in Borneo Tropical Rainforests Associated with the 1983 El Niño Southern Oscillation Event." In G. T. Prance (ed.), *Tropical Forests and the World Atmosphere*, 75–102. AAAS Symposium, Washington, DC.

Peters, C. M, A. H. Gentry, and R. Mendelsohn. 1989. "Valuation of a Tropical Forest in Peruvian Amazonia." *Nature* 339:655–56.

CHAPTER TWO

Allen, J., and D. Barnes. 1985. "The Causes of Deforestation." *Annals of the Association of American Geographers*. Clark University, Worcester, MA.

Brazier, J. D. 1982. "Patterns, Trends and Forecasts of Wood Consumption to the Year 2000." Address to the Annual Meeting of the British Association for the Advancement of Science. September.

Burns, D. 1986. "Runway and Treadmill Deforestation: Reflections on the Economics of Forest Development in the Tropics." Series Paper on Tropical Forest Policy Paper No. 2. Forest and Land Use Programme, International Union for Conservation of Nature and Natural Resources/International Institute for Environment and Development (IUCN/IIED), London.

Cardellichio, P. A., and D. M. Adams. 1988. "Evaluation of the IIASA Model of the Global Forest Sector." *CINTRAFOR Working Paper 13*. Center for International Trade in Forest Products, University of Washington, Seattle, WA.

Cardellichio, P.A., Y. C. Youn, C. S. Binkley, J. R. Vincent, and D. M. Adams. 1988. "An Economic Analysis of Short-run Timber and Timber Supply around the Globe." *CINTRAFOR Working Paper 18*. Center for International Trade in Forest Products, University of Washington, Seattle, WA.

Cardellichio, P. A., Y. C. Youn, D. M. Adams, R. W. Joo, and J. T. Chmelik. 1989. "A Preliminary Analysis of Timber and Timber Products Production, Consumption, Trade, and Prices in the Pacific Rim until 2000." *CIN-*

TRAFOR Working Paper 22. Center for International Trade in Forest Products, University of Washington, Seattle, WA.

Caufield, C. 1985. *In the Rainforest: Report from a Strange, Beautiful, Imperiled World.* Alfred A. Knopf, New York, NY.

Davidson, J. 1985. *Economic Use of Tropical Moist Forests.* International Union for Conservation and of Nature and Natural Resources (IUCN), Gland, Switzerland.

Dickinson, R. E. 1981. "Effects of Tropical Deforestation on Climate." *Studies in Third World Societies* 14:11–442.

Durning, A. 1989. "Mobilizing at the Grassroots." In *State of the World 1989.* Worldwatch Institute, Washington, DC.

Eckholm, E. 1982. *Down to Earth: Environment and Human Needs.* International Institute for Environment and Development, Norton, New York, NY.

Erfuth, T. 1984. "Trends in Timber Supplies from Tropical Regions." In: Proceedings 7341: International Forest Products Trade, Forest Products Research Society, Madison, WI.

Food and Agriculture Organization of the United Nations (FAO) 1981a. *Forest Resources of Tropical America: Part II: Country Briefs.* FAO, Rome.

————. 1981b. *Forest Resources of Tropical Africa.* FAO, Rome.

————. 1981c. *Forest Resources of Tropical Asia.* FAO, Rome.

————. 1985. *Tropical Forest Action Plan.* FAO, Rome.

————. 1987. ITTO Preliminary Market Transparency Study for Tropical Timber Production in the ASEAN Region. ITTO, Yokahama, Japan.

————. 1988. *Forest Products: World Outlook Projections.* FAO Forestry Paper 84. FAO, Rome.

Gammie, J. I. 1981. "World Timber to the Year 2000." Special Report No. 98. Economist Intelligence Unit, London.

Grainger, A. 1986. "The Future Role of the Tropical Rain Forests in the World Forest Economy." Ph.D. Dissertation. Oxford University, St. Cross College, Oxford.

————. 1987a. "The Future Environment for Forest Management in Latin America." In *Management of the Forests of Tropical America: Prospects and Technologies.* Institute of Tropical Forestry/USDA Forest Service, Washington, DC.

————. 1987b. "A Land Use Simulation Model for the Humid Tropics." Paper presented at the Land and Resource Evaluation for National Planning Conference and Workshop. Chetumal, Mexico, January 25–31.

————. 1988. "Future Supplies of High-Grade Tropical Hardwoods from Intensive Plantations." *Journal of World Forest Resource Management* 3:15–29.

————. In press. *The Tropical Rain Forests and Man.* Columbia University Press, New York, NY.

Kallio, M., D. P. Dykstra and C. S. Binkley (eds.). 1987. *The Global Forest Sector: An Analytical Perspective.* Wiley & Sons, Chichester, England.

Kauman, W. 1987. "Prospective Markets for Tropical Forest Products." In *Management of the Forests of Tropical America: Prospects and Technologies.* Institute of Tropical Forestry/USDA Forest Service, Washington, DC.

Keipi, K. 1987. "Tropical Forest Management in Latin America: Role of the InterAmerican Development Bank." In *Management of the Forests of Tropical America: Prospects and Technologies*. Institute of Tropical Forestry, USDA Forest Service.

Kunkle, S. 1978. "Forestry Support for Agriculture Through Watershed Management, Windbreaks and Other Conservation Actions." Presented to Eighth World Forestry Congress. Jakarta, October 16–28.

Laarman, J. G. 1988. "Export of Tropical Hardwoods in the Twentieth Century." In J. F. Richards and R. P. Tucker (eds.), *World Deforestation in the Twentieth Century*. Duke University Press, Durham, NC.

Lanly, J. 1982. *Tropical Forest Resources*. FAO Forestry Paper No. 30. FAO, Rome.

Leonard, J. 1987. "Natural Resources and Economic Development in Central America: A Regional Environmental Profile." Draft prepared for the International Institute for Environment and Development, and Regional Office for Central America. U.S. Agency for International Development (USAID), Washington, DC.

Leonard, J., and J. Nations. 1986. "Grounds of Conflict in Central America." In A. Maguire and J. Brown (eds.), *Bordering on Trouble: Resources and Politics in Latin America*. Adler and Adler, Bethesda, MD.

Myers, N. 1980. *Conversion of Tropical Moist Forests*. National Academy of Sciences, Washington, DC.

————. 1984. *The Primary Source: Tropical Forests and Our Future*. W.W. Norton and Co., New York and London.

————. 1987. *Not Far Afield: U.S. Interests and the Global Environment*. World Resources Institute, Washington, DC.

Repetto, R. 1987. "Creating Incentives for Sustainable Forest Development." *Ambio* 16(2–3):94–99.

Repetto, R., and M. Gillis (eds.). 1988. *Public Policies and the Misuse of Forest Resources*. Cambridge University Press, NY.

Richardson, D. 1970. The End of Forestry in Great Britain. In *Advancement of Science* (December).

Schmink, M. 1987. "The Rationality of Tropical Forest Destruction." In *Management of the Forests of Tropical America: Prospects and Technologies*. Institute of Tropical Forestry/USDA Forest Service, Washington, DC.

Scott-Kemis, D. 1982. *Transnational Corporations and Tropical Industrial Forests*. Science Policy Research Unit, University of Sussex.

Spears, J. 1982. "Preserving Watershed Environments." *Unasylva* 34:137.

U.S. Interagency Task Force on Tropical Forests. 1980. *The World's Tropical Forests: A Policy, Strategy and Program for the United States*. Report to the President. Government Printing Office, Washington, DC.

Vietmeyer, N. 1975. *Underexploited Tropical Plants with Promising Economic Value*. National Academy of Science, Washington, DC.

Vincent, J. R. 1990. "Rent Capture and the Feasibility of Tropical Forest Management." *Land Economics* 66:212–22.

————. 1991. "Tropical Timber Trade, Industrialization, and Policies"

Appendix 2: "Forecasts of Wood Products Production, Consumption, Trade and Prices." Unpublished. Harvard Institute for International Development, Cambridge, MA.

Vincent, J. R., and Y. Hadi. 1991. "Deforestation and Agricultural Expansion in Peninsular Malaysia." Development Discussion Paper. Harvard Institute for International Development, Cambridge, MA.

Woodwell, G. M., J. E. Hobbie, R. A. Houghton, J. M. Melillo, B. Moore, B. J. Peterson, and E. R. Shaver. 1983. "Global Deforestation: Contribution to Atmospheric Carbon Dioxide." *Science* 222:1081–86.

World Bank. 1978. *Forestry Sector Policy Paper*. Washington, DC.

———. 1980. *Energy in Developing Countries*. Washington, DC.

———. 1989. *Price Prospects for Major Primary Commodities, 1988–2000*. Washington, DC.

World Resources Institute (WRI). 1986. *World Resources 1986*. World Resources Institute and International Institute for Environment and Development. Basic Books, New York, NY.

———. 1987. *World Resources 1987*. International Institute for Environment and Development and World Resources Institute, Basic Books, New York, NY.

———. 1988. *World Resources 1988*. International Institute for Environment and Development and World Resources Institute, Basic Books, New York, NY.

1990. *World Resources 1990–91*. In collaboration with The United Nations Environment Programme and The United Nations Development Programme. Oxford University Press, NY.

World Wildlife Fund (WWF). 1987. "Tropical Forest Conservation and the ITTA." WWF Position Paper 2. Gland, Switzerland.

Zobel, B. 1984. "The Changing Quality of the World Wood Supply." *Wood Science Technology* 18(1):1.

CHAPTER THREE

Appanah, S., and Salleh Mohd. Nor. 1987. "Natural Regeneration and Its Implications for Forest Management in the Malaysian Dipterocarp Forests." Draft Manuscript.

Asabere, P. K. 1987. "Attempts at Sustained Yield Management in the Tropical High Forests of Ghana." In F. Mergen and J. R. Vincent (eds.), *Natural Management of Tropical Moist Forest*, 47–70. Yale University School of Forestry and Environmental Studies, New Haven, CT.

Ashton, P. M. S. 1992. "Establishment and Early Growth of Advance Regeneration of Canopy Trees in Moist Mixed-Species Forest." In M. D. Kelty, B. C. Larson, and C. D. Oliver (eds.), *The Ecology and Management of Mixed Species Stands*, 101–22. Kluwer Academic Press, Dortrecht, the Netherlands.

Ashton, P. M. S., C. V. S. Gunatilleke, and I. A. N. Gunatilleke. In press. "A

Shelterwood Method for Sustained Timber Production in *Mesua-Shorea* Forest of Southwest Sri Lanka." In E. Erdelen, Ch. Preu, N. Ishwaran, and Ch. Santiapillai (eds.), *Ecology and Landscape Management in Sri Lanka; Conflict or Compromise?* Joseph Margraf Scientific, Hamburg, Germany.

Ashton, P. S. 1982. "Dipterocarpaceae." In C.G.G.J. van Steenis (ed.), *Flora Malesiana* 9: 237–552.

Bazzaz, F. A. 1983. "Characteristics of Populations in Relation to Disturbance in Natural and Man-modified Ecosystems." In H. A. Mooney and M. Godron (eds.), *Disturbance and Ecosystems—Components and Response*, 259–75. Springer-Verlag, New York, NY.

————. 1984. "Dynamics of Wet Tropical Forests and Their Species Strategies." In E. Medina, H. A. Mooney, and C. Vazquez-Yanes (eds.), *Physiological Ecology of Plants of the Wet Tropics*, 233–43. D. W. Junk, The Hague.

Bazzaz, F. A., and S. T. A. Pickett. 1980. "Physiological Ecology of Tropical Succession: A Comparative Review." *Annual Review of Ecology and Systematics* 11:287–310.

Becker, P. 1983. "Seedling Biology of Two Closely Related *Shorea* Species (Dipterocarpaceae)." Ph.D. Dissertation, University of Michigan, Ann Arbor, MI.

Bjorkman, O., and M. M. Ludlow. 1972. "Characterization of the Light Climate on the Floor of a Queensland Rain Forest." *Carnegie Institute of Washington Yearbook* 71:85–94.

Bowes, M. D., and J. V. Krutilla. 1989. *Multiple-Use Management: The Economics of Public Forestlands*. Resources for the Future, Washington, DC.

Brokaw, N. V. L. 1982. "Treefalls." In E. G. Leigh, Jr., A. S. Rand, and D. M. Windsor (eds.), *The Ecology of a Tropical Forest: Seasonal Rhythms and Long-Term Changes*. Smithsonian Institution Press, Washington, DC.

————. 1985. "Treefalls, Regrowth, and Community Structure in Tropical Forests." In S. T. A. Pickett and P. S. White (eds.), *The Ecology of Natural Disturbance and Patch Dynamics*. Academic Press, New York, NY.

Brown, W. H., and D. M. Mathews. 1914. "Philippine Dipterocarp Forests." *The Philippine Journal of Science* 9(5):413–561.

Burgess, P. F. 1968. "An Ecological Study of the Hill Forests of the Malay Peninsula." *Malaysian Forester* 31(4):314–25.

Burns, D. 1986. "Runway and Treadmill Deforestation: Reflections on the Economics of Forest Development in the Tropics." Series Paper on Tropical Forest Policy No. 2. Forestry and Land Use Programme. International Union for Conservation of Nature and Natural Resources/International Institute for Environment and Development (IUCN/IIED), London.

Chazdon, R. L. 1985. "Leaf Display, Canopy Structure, and Light Interception of Two Understory Palm Species." *American Journal of Botany* 72:1493–1502.

Chazdon, R. L., and N. Fetcher. 1984. "Photosynthetic Light Environments in a Lowland Rain Forest in Costa Rica." *Journal of Ecology* 72:553–64.

de Graf, N. R. 1986. *A Sylvicultural System for Natural Regeneration of Tropical Rain Forest in Suriname*. Agricultural University, Wageningen, the Netherlands.

Denslow, J. S. 1980. "Gap Partitioning Among Tropical Rain Forest Trees." *Biotropica* 12(suppl.):47–55.

Food and Agriculture Organization of the United Nations (FAO). 1981. *Forest Resources of Tropical Asia.* UN32/6.1301-78-041 Technical Report 3, with UNDP. FAO, Rome.

Garwood, N. C. 1983. "Seed Germination in a Seasonal Tropical Forest in Panama: A Community Study." *Ecological Monographs* 53:159–81.

Gunatilleke, C. V. S., and P. S. Ashton. 1987. "New Light on the Plant Geography of Ceylon II: The Ecological Biogeography of the Lowland Endemic Tree Flora." *Journal of Biogeography* 14:295–327.

Hartshorn, G. S. 1978. "Tree Falls and Tropical Forest Dynamics." In P. B. Tomlinson and M. H. Zimmerman (eds.), *Tropical Trees as Living Systems*, 617–38. Cambridge University Press, New York, NY.

———. 1980. "Neotropical Forest Dynamics." *Biotropica* 12:23–30.

Hartshorn, G. S., R. Simeone, and J. A. Tosi, Jr. 1987. "Sustained Yield Management of Natural Forests: A Synopsis of the Palcazu Development Project in the Central Selva of the Peruvian Amazon." Unpublished. Tropical Science Center, Costa Rica.

Hubbell, S. P. 1990. "The Fate of Juvenile Trees in Neotropical Forest: Implications for the Natural Maintenance of Tropical Tree Diversity." In K. S. Bawa and M. Hadley (eds.), *Reproductive Ecology of Tropical Forest Plants*, 317–41. Parthenon Publishing, Park Ridge, NY.

Hubbell, S. P., and R. B. Foster. 1983. "Diversity of Canopy Trees in a Neotropical Forest and Implications for Conservation." In S. Sutton, A. Chadwick, and T. C. Whitmore (eds.), *The Tropical Rain Forest*, 25–41. Blackwell Scientific Publications, Oxford.

Hutchinson, I. D. 1979. *Liberation Thinning: A Tool in the Management of Mixed Dipterocarp Forest in Sarawak.* UNDP/FAO/MAL/76/008, Sarawak Forestry Department, Kuching, Sarawak, Malaysia.

———. 1982. *Forestry Development Project Sarawak: Terminal Report.* UNDP/FAO/MAL/76/008, Sarawak Forestry Department, Kuching, Sarawak, Malaysia.

———. 1987a. "Improvement Thinning in Natural Tropical Forests: Aspects and Institutionalization." In F. Mergen and J. R. Vincent (eds.), *Natural Management of Tropical Moist Forest*, 113–34. Yale University School of Forestry and Environmental Studies, New Haven, CT.

———. 1987b. "The Management of Humid Tropical Forest to Produce Wood." In *Management of the Forests of Tropical America: Prospects and Technologies*, pp. 121–55. Proceedings of a conference sponsored by USAID, USDA Forest Service, and Office of International Cooperation and Development, Puerto Rico, September 22–26.

Ismail b. Hj. Ali. 1966. "A Critical Review of Malayan Silviculture in the Light of Changing Demand and Form of Timber Utilization." *Malaysian Forester* 29(4):228–33.

Jabil, M. 1983. "Problems and Prospects in Tropical Rain Forest Management for Sustained Yield." *Malaysian Forester* 46(4):398–408.

Jonkers, W. B. J., and J. Hendrison. 1987. "Prospects for Sustained Yield Management of Tropical Rain Forest in Surinam." In *Management of the Forests of Tropical America: Prospects and Technologies*, 157–73. Proceedings of a conference sponsored by USAID, USDA Forest Service, and Office of International Cooperation and Development, Puerto Rico, September 22–26, 1986. Institute of Tropical Forestry, USDA Forest Service.

Jordan, C. F. 1982. "Amazon Rain Forests." *American Scientist* 70(4):394–401.

Kio, P. R. O. 1979. "Management Strategies in the Natural Tropical High Forest." *Forest Ecology and Management* 2:207–20.

———. 1987. "Perspective for Africa." Consultancy Report for Harvard Institute for International Development (HIID), Cambridge, MA.

Kio, P. R. O., and S. A. Ekwebelam. 1987. "Plantations Versus Natural Forests for Meeting Nigeria's Wood Needs." In F. Mergen and J. R. Vincent (eds.), *Natural Management of Tropical Moist Forest*, 149–76. Yale University School of Forestry and Environmental Studies, New Haven, CT.

Kochummen, K. M., and F. S. P. Ng. 1977. "Natural Plant Succession After Farming at Kepong." *Malaysian Forester* 40(1):61–78.

Korsgaard, S. 1986. "An Analysis of the Potential for Timber Production Under Conservation Management in the Tropical Rain Forest of South East Asia." Unpublished report to the Danish Government. Research Council for Development Research, Copenhagen, Denmark.

Lanly, J. 1982. *Tropical Forest Resources*. FAO Forestry Paper No. 30. FAO, Rome.

Leslie, A. 1977. "Where Contradictory Theory and Practice Co-exist." *Unasylva* 29(115):2–17.

———. 1980. "Logging Concessions: How to Stop Losing Money." *Unasylva* 32(129):2–8.

———. 1987. "A Second Look at the Economics of Natural Management Systems in Tropical Mixed Forests." *Unasylva* 39(155):46–58.

Lowe, R. G. 1978. "Experience with the Tropical Shelterwood System of Regeneration in Natural Forest in Nigeria." *Forest Ecology and Management* 1:193–212.

Lowe, R. G., and F. N. Ugbechie. 1975. "Kennedy's Natural Regeneration Experiments at Sapoba, After 45 Years." Unpublished report to the Forestry Association of Nigeria, 6th Annual Conference. Calabar, Nigeria.

Marn, H. M., and W. B. Jonkers. 1981. "Logging Damage in Tropical High Forest." Paper presented at the International Forestry Seminar, Kuala Lumpur, November 1980. Sarawak Forestry Department, Kuching, Sarawak, Malaysia.

Myers, N. 1980. "The Present Status and Future Prospects of Tropical Moist Forest." *Environmental Conservation* 7(2):101–14.

National Research Council (NRC). 1982. *Ecological Aspects of Development in the Humid Tropics*. National Academy Press, Washington, DC.

Ng, F. S. P. 1980. "Germination Ecology of Malaysian Woody Plants." *Malaysian Forester* 63:606–37.

Nwoboshi, C. L. 1987. "Regeneration Success of Natural Management, Enrichment Planting, and Plantations of Native Species in West Africa." In

F. Mergen and J. R. Vincent (eds.), *Natural Management of Tropical Moist Forest*, 71–92. Yale University School of Forestry and Environmental Studies, New Haven, CT.

Office of Technology Assessment (OTA), United States Congress. 1984. *Technologies to Sustain Tropical Forest Resources*. U.S. Government Printing Office, Washington, DC.

Pearcy, R. W. 1983. "The Light Environment and Growth of C3 and C4 Tree Species in the Understory of a Hawaiian Forest." *Oecologia* 58:26–32.

Poore, D. P., P. Burgess, J. Palmer, S. Rietbergen, and T. Synnott. 1989. *No Timber without Trees*. Earthscan, London.

Richards, P. W. 1952. *The Tropical Rain Forest: An Ecological Study*. Cambridge University Press, London.

Schmidt, R. 1987. "Tropical Rain Forest Management: A Status Report." *Unasylva* 39(156):2–17.

Strong, D. R. 1977. "Epiphyte Loads, Treefalls, and Perennial Forest Disruption: A Mechanism for Maintaining Higher Tree Species Richness in the Tropics without Animals." *Journal of Biogeography* 4:215–18.

Tang, H. T. 1974. "A Brief Assessment of the Regeneration Systems for Hill Forests in Peninsular Malaysia." *Malaysian Forester* 37(4):263–70.

————. 1987. "Problems and Strategies for Regenerating Dipterocarp Forests in Malaysia." In F. Mergen and J. R. Vincent (eds.), *Natural Management of Tropical Moist Forest*, 23–46. Yale University, School of Forestry and Environmental Studies, New Haven, CT.

Tosi, J. A., Jr. 1982. *Sustained Yield Management of Natural Forests*. Report prepared for the Office of Development Resources, USAID/Peru. Tropical Science Center, Costa Rica.

Uebelhör, K., B. Lagundino, and R. Abalus. 1990. *Appraisal of the Philippine Selective Logging System*. GTZ Technical Report 7. Philippine-German Dipterocarp Forest Management Project PN88.2047.4.

Vincent, J. R., J. Chamberlain, and S. Warren. 1987. "Summary." In F. Mergen and J. R. Vincent (eds.), *Natural Management of Tropical Moist Forest*, 200–203. Yale University School of Forestry and Environmental Studies, New Haven, CT.

Wadsworth, F. H. 1987. "Applicability of Asian and African Silviculture Systems to Naturally Regenerated Forests of the Neotropics." In F. Mergen and J. R. Vincent (eds.), *Natural Management of Tropical Moist Forest*, 93–112. Yale University School of Forestry and Environmental Studies, New Haven, CT.

Whitmore, T. C. 1974. "Change with Time and the Role of Cyclones in Tropical Rainforest on Kolombangara, Solomon Islands." *Commonwealth Forestry Institute Paper No. 46*, Oxford University, Oxford.

————. 1984. *Tropical Rain Forests of the Far East* (2d edition). Clarendon Press, Oxford.

Wyatt-Smith, J. 1963. "Manual of Malayan Silviculture for Inland Forests." *Malaysian Forest Records* No. 23. Forestry Department, Kuala Lumpur, Malaysia.

_____. 1987a. "The Management of Tropical Moist Forest for the Sustained Production of Timber: Some Issues." Series Paper on Tropical Forest Policy. Forestry and Land Use Programme. International Union for Conservation of Nature and Natural Resources/International Institute for Environment and Development (IUCN/IIED), London.

_____. 1987b. "Problems and Prospects for Natural Management of Tropical Moist Forest." In F. Mergen and J. R. Vincent (eds.), *Natural Management of Tropical Moist Forest*. Yale University, School of Forestry and Environmental Studies, New Haven, CT.

Yap, S. K. 1982. "The Reproductive Biology of Some Understory Species in the Lowland Dipterocarp Forest of West Malaysia." Ph.D. Dissertation, University of Malaya, Kuala Lumpur.

CHAPTER FOUR

Boado, E. K. 1988. "Incentive Policies and Forest Use in the Philippines." In R. Repetto and M. Gillis (eds.), *Public Policies and the Misuse of Forest Resources*, 165–203. Cambridge University Press, New York, NY.

Bowes, M. D., and J. V. Krutilla. 1989. *Multiple-Use Management: The Economics of Public Forestlands*. Resources for the Future, Washington, DC.

Browder, J. 1988. "Public Policy and Deforestation in the Brazilian Amazon." In R. Repetto and M. Gillis (eds.), *Public Policies and the Misuse of Forest Resources*, 247–97. Cambridge University Press, New York, NY.

Gillis, M. 1980. "Fiscal and Financial Issues in Tropical Hardwood Concessions." *Development Discussion Paper No. 110*. Harvard Institute of International Development, Cambridge, MA.

_____. 1988a. "Indonesia: Public Policies, Resource Management and the Tropical Forest." In R. Repetto and M. Gillis (eds.), *Public Policies and the Misuse of Forest Resources*, 43–113. Cambridge University Press, New York, NY.

_____. 1988b. "Malaysia: Public Policies and the Tropical Forest." In R. Repetto and M. Gillis (eds.), *Public Policies and the Misuse of Forest Resources*, 115–203. Cambridge University Press, New York, NY.

Repetto, R. 1987. "Creating Incentives for Sustainable Forests in Indonesia." *Ambio*. 16:94–99.

_____. 1988. "Economic Policy Reform for Natural Resource Conservation." World Bank, Washington, DC.

Repetto, R., and M. Gillis (eds.). 1988. *Public Policies and the Misuse of Forest Resources*. Cambridge University Press, New York, NY.

Vincent, J. R. 1992. "Forest Sector Models in Developing Countries." Invited paper for symposium on Forest Sector Trade and Environmental Impacts Models, April 30 to May 1, 1992, College of Forest Resources, University of Washington, Seattle (in press).

Vincent, J. R., and C. S. Binkley. 1991. "Forest-based Industrialization: A Dynamic Perspective." *Development Discussion Paper No. 389*. Harvard Institute for International Development, Cambridge, MA.

CHAPTER FIVE

Afolayan, T. A. 1980. "A Synopsis of Wildlife Conservation in Nigeria." *Environmental Conservation* 7(3):207–12.

Ajayi, S. S. 1979. *Utilization of Forest Wildlife in West Africa.* FAO:MISC/79/26. FAO, Rome.

Allegretti, M. H., and S. Schwartzman. 1986. *Extractive Resources: A Sustainable Development Alternative for Amazonia.* Mimeograph Report to World Wildlife Fund, Washington, DC.

Anderson, A. B. 1983. "The Biology of *Orbignya martiana*, Palmae, a Tropical Dry Forest Dominat in Brazil." Ph.D. Thesis, University of Florida. University Microfilms International, Ann Arbor, MI.

Anderson, A. B., P. H. May, and M. J. Balick. 1991. *The Subsidy from Nature, Palm Forests, Peasantry, and Development on an Amazon Frontier.* Columbia University Press, New York, NY.

Anonymous. 1990. "The Promotion of Sustainable Forest Management: A Case Study in Sarawak, Malaysia." A report submitted to the International Tropical Timber Council by provision established persuant to resolution I(VI). Unpublished report.

Aruga, J. A., and S. M. Saulei. 1990. "The Status and Prospects of Non-Timber Product Utilization in Papua New Guinea." Seminar on Sustainable Development of Tropical Forests (draft report), Kamakura, Japan. Permanent Committee on Reforestation and Forest Management, Quito, Ecuador.

Ashton, P. S., T. J. Givnish, and S. Appanah. 1988. "Staggered Flowering in the Dipterocarpaceae: New Insights into Floral Induction and the Evolution of Mast Fruiting in the Seasonal Tropics." *The American Naturalist* 132(1):44–66.

Balick, M. J., and R. Mendelsohn. 1992. "Assessing the Economic Value of Traditional Medicines from Tropical Rainforests." *Conservation Biology* 6(1):In press.

Brennan, J. 1981. *The Original Thai Cookbook.* Richard Marek Publishers, Inc., New York, NY.

Burkill, I. H. 1966. *A Dictionary of the Economic Products of the Malay Peninsula.* Governments of Malaysia and Singapore, Ministry of Agriculture and Co-operatives.

Caldecott, J. 1987. *Hunting and Wildlife Management in Sarawak.* World Wildlife Fund, Malaysia.

Clay, Jason W. 1988. "Indigenous Peoples and Tropical Forest: Models of Land Use and Management from Latin America." *Cultural Survival* Report No. 27.

Cody, D. 1983. *Manual of the Production of Rattan Furniture.* United Nations, New York, NY.

Collins, N. M., and M. G. Morris. 1985. "Threatened Swallowtail Butterflies of the World." *The IUCN Red Data Book.* International Union for the Conservation of Nature (IUCN), Gland and Cambridge, England.

Consultative Group on Biological Diversity (CGBD). 1989. *Marketing Non-Timber Tropical Forest Products: Prospects and Promise.* A Workshop Report. *Cultural Survival,* Cambridge, MA.

de Beer, J. 1990. "Subsistence Use and Market Value of Non-Timber Forest Products: The Example from Southeast Asia." Seminar on Sustainable Development of Tropical Forests (Draft Report), Kamakura, Japan. Permanent Committee on Reforestation and Forest Management, Quito, Ecuador.

de Beer, J., and M. McDermott. 1989. "The Economic Value of Non-timber Forest Products in Southeast Asia." Netherlands Committee for the International Union for Conservation of Nature and Natural Resources (IUCN), Amsterdam.

de Vos, A. 1977. "Game as Food: A Report on Its Significance in Africa and Latin America." *Unasylva* 29(116).

Dixon, J. A., and P. B. Sherman. 1990. *Economics of Protected Areas.* East-West Center. Island Press, Washington, DC.

Dourojeanni, M. 1985. "Over-exploited and Under-used Animals in the Amazon Region." In G. Prance and T. Lovejoy (eds.), *Amazonia.* Pergamon Press, Oxford.

Dransfield, J. 1977a. "The Identity of 'Rotan Manau' in the Malay Peninsula." *Malaysian Forester* 40:197–99.

_____. 1977b. "*Calamus caesius* and *Calamus trachycoleus* Compared." *Gard. Bull. Sing.* 30:75–78.

_____. 1988. "The Identity of Sika in Palawan, Philippines." *Kalikasan: Philippine Journal of Biology* 9(1):43–48.

_____. 1981. "The Biology of Asiatic Rattans in Relation to the Rattan Trade and Conservation." In H. Synge (ed.), *The Biological Aspects of Rare Plant Conservation,* 179–86. John Wiley and Sons, New York, NY.

_____. 1987a. "Prospects for Rattan Cultivation." Unpublished Manuscript.

_____. 1987b. "The Conservation Status of Rattan in 1987: A Cause for Great Concern." Paper delivered at the International Rattan Seminar, November 12–14, 1987. Chiangmai, Thailand.

Drostitz, W. 1979. "The New International Market for Game Meat." *Unasylva* 31(123).

Ferreyra. 1959. "Camau-Camu, Nueva Fuenta Nacional de Vitamina C." *Bol. Exp. Agropecuaria* 7(4):28.

Food and Agriculture Organization of the United Nations (FAO). 1969. "The Role of Wildlife and National Parks in Tropical Forestry." Secretariat Note FAO Committee on Forest Development in the Tropics. Rome, Italy. October 21–24, 1964.

_____. 1980. *1979 Yearbook of Fishery Statistics: Catches and Landings.* FAO Fisheries Series 48. FAO, Rome.

Godoy, R., and C. P. A. Bennett. 1987. "Monocropped and Intercropped Coconuts in Indonesia: Project Goals Which Conflict with Smallholder Interests." Unpublished Manuscript.

Godoy, R., and R. Lubowski. 1991. "How Much Is the Forest Worth? Guidelines for the Valuation of Non-Timber Tropical Forest Products." Unpublished Manuscript. Harvard Institute for International Development, Cambridge, MA.

Godoy, R., and D. Rodrik. 1990. "The Costs and Consequences of Export Restrictions on Non-timber Tropical Forest Products: Indonesia's Export Policy on Rattan." Unpublished Manuscript. Harvard Institute for International Development, Cambridge, MA.

Godoy, R., and C. F. Tan. 1989. "The Profitability of Smallholder Rattan Cultivation in Southern Borneo, Indonesia." *Human Ecology* 17(3):347–63.

Godoy, R., and D. Wilkie. 1991. "The Effect of Economic Development on Non-Timber Tropical Forest Product Extraction." Draft Report. Harvard University, Cambridge, MA, and Tufts University, Medford, MA.

Government of Thailand. 1988. "Foreign Trade Statistics." Supplied by the Department of Customs.

Hayes, W. B. 1953. *Fruit Growing in India*. Kitabistan, Allahabad.

Heyne, K. 1927. *De Nuttige Planten van Nederlandsch Indië*. 2 volumes. *Het Departement van Landbouw, Niguer & Handel, Nederlansch, Indië*.

Inskipp, T., and S. Wells. 1979. *International Trade in Wildlife*. Earthscan, International Institute for Environment and Development, Fauna Preservation Society.

Institute for Amazonian Studies (IEA). 1989. *Man and the Environment in Amazonia, Potential Forest Use and the Social Management of Natural Resources*. Curitiba, Brazil.

International Development Research Center (IDRC). 1980. "Rattan: A Report of a Workshop Held in Singapore." International Development Research Center, Ottawa.

Janzen, D. H. 1974. "Tropical Blackwater Rivers, Animals, and Mast Fruiting by the Dipterocarpaceae." *Biotropica* 6:69–103.

Kahn, F. 1988. "Ecology of Economically Important Palms in Peruvian Amazonia." *Advances in Economic Botany* 16:42–49.

Kio, P. R. O. 1987. "Perspective for Africa." Consultancy Report for the Harvard Institute for International Development, Cambridge, MA.

Krostitz, W. 1979. "The New International Market for Game Meat." *Unasylva* 31(123).

Leighton, M. 1987. "Rattan Production and Industry in West Kalimantan." Unpublished Manuscript. Harvard Institute for International Development, Cambridge, MA.

Lessard, G., and A. Chouinard (eds.). 1980. *Bamboo Research in Asia*. Proceedings of a workshop held in Singapore, 22–30 May 1980, International Development Research Center (Ottawa).

Macmillan, H. F. 1991. *Tropical Planting and Gardening* (6th edition). Malayan Nature Society, Kuala Lumpur.

Malhotra, K. D., and M. Poffenberger (eds.). 1989. *Forest Regeneration through Community Protection: The West Bengal Experience*. Proceedings of the Working

Group Meeting on Forest Protection Committees, Calcutta, 21–22 June 1989. West Bengal Forest Department, West Bengal.

Manokaran, N. 1981. "Clustering in Rotan Manau (*Calamus manan*)." *Malaysian Forester* 44:557–60.

_____. 1984. "Biological and Ecological Considerations Pertinent to the Silviculture of Rattans." *Proceedings of the Rattan Seminar*, Kuala Lampur, Malaysia.

Manokaran, N., and K. M. Wong. 1983. "The Silviculture of Rattans—An Overview with Emphasis on Experiences from Malaysia." *Malaysian Forester* 46(3).

Marn, H. M., and W. B. Jonkers. 1981. "Logging Damage in Tropical High Forest." Paper presented at the International Forestry Seminar, Kuala Lumpur, November 1980. Sarawak Forestry Department, Kuching, Sarawak, Malaysia.

Mei, L., and L. Sieh. 1985. *International Marketing: Cases from Malaysia.* Institute of Southeast Asian Studies, Singapore.

Menon, K. D. 1980. *Rattan*: A Report of a Workshop Held in Singapore. International Development Research Center, Ottawa.

Meulenhoff, L. W., and T. M. Silitonga. 1978. "The Importance of Minor Forest Products." In *Proceedings from Eighth World Forestry Congress in Jakarta, Indonesia.*

Moore, H. E. 1973. "The Major Groups of Palms and Their Distribution." *Gentes Herb.* 11(2):27–141.

Murray, G. F. 1988. "The Wood Tree as a Peasant Cash Crop: An Anthropological Strategy for the Domestication of Energy (Haiti)." In L. Fortmann and J. Bruce (eds.), *Whose Trees? Proprietary Dimensions of Forestry.* Westview Press, Boulder, CO.

Myers, G. S. 1947. "The Amazon and Its Fishes." *Aquarium Journal* 18(4):13–20.

Myers, N. 1984. *The Primary Source: Tropical Forests and Our Future.* W.W. Norton and Company, New York, NY.

_____. 1986. "Tropical Forest: Patterns of Depletion" in G. T. Prance (ed.), *Tropical Forests and the World Atmosphere.* American Association for the Advancement of Science (AAAS) Symposium, Washington, DC.

National Academy of Science (NAS). 1980. *Firewood Crops: Shrubs and Tree Species for Energy Production.* National Research Council Press, Washington, DC.

Noor, S., and W. Razali. 1987. "The Growth and Yield of a Nine-Year-Old Rattan Plantation." Paper presented at the International Rattan Seminar, November 12–14, 1987, Chiangmai, Thailand.

Ntiamoa-Baidu, Y. 1987. "West African Wildlife: A Resource in Jeopardy." *Unasylva* 39(156).

Obiorah, B. A. 1986. "Sourcing Pharmaceutical Raw Materials from Indigenous Medicinal Plants." In A. Sofowara (ed.), *The State of Medicinal Plants Research in Nigeria*, 79–88. Nigerian Society of Pharmacognosy.

Ochse, J. et al. 1961. *Tropical and Subtropical Agriculture*. Macmillan, New York, NY.

Okafor, J. C. 1979. "Edible Indigenous Woody Plants in the Rural Economy of the Nigeria Forest Zone." In D. U. U. Okali (ed.), *The Nigerian Rainforest Ecosystem*.

Peters, C. M., A. H. Gentry, and R. Mendelsohn. 1989. "Valuation of a Tropical Forest in Peruvian Amazonia." *Nature* 339:655–56.

Pierret, P. V., and M. Dourojeanni. 1966. "La Caza y la Alimentacion Humana en las Riberas del rio Pachitea." Peru, *Turrialba* 16(3).

Prance, G. 1984. "The Use of Edible Fungi by Amazonian Indians." *Advances in Economic Botany* 1:127–39.

Prescott-Allen, R. 1982. *What's Wildlife Worth? Economic Contributions of Wild Plants and Animals to Developing Countries*. International Institute for Environment and Development (IIED), London and Washington, DC.

Purseglove, J. W. 1972. *Tropical Crops: Monocotyledons*. Longman, London.

———. 1968. *Tropical Crops: Dicotyledons*. J. Wiley, New York, NY.

Repetto, R., and M. Gillis (eds.). 1988. *Public Policies and the Misuse of Forest Resources*. Cambridge University Press, New York, NY.

Repetto, R., W. Magrath, M. Wells, C. Beer, and F. Rossini. 1989. *Wasting Assets: Natural Resources in the National Income Accounts*. World Resources Institute, Washington, DC.

Roca, N. 1965. "Estudios Guimico—Bromatologica de la *Myrciaria paraensis*." *Berg. Tesis* Guimica, Univ. Nac. Mayor San Marcos, Lima.

Rosengarten, F. R. 1984. *The Book of Edible Nuts*. Walker and Company, New York, NY.

Sarawak Forestry Department. *Annual Reports for the Forest Department*, 1970–1980. Kuching, Sarawak, Malaysia.

Sayer, J. A. 1990. "Using Non-timber Products to Support Forest Conservation Programmes." Seminar on Sustainable Development of Tropical Forests (Draft Report), Kamakura, Japan. Permanent Committee on Reforestation and Forest Management, Quito, Ecuador.

Sharma, Y. M. L. 1980. "Bamboo in the Asia-Pacific Region." In Lessard and Chouinard (eds.), *Bamboo Research in Asia: Proceedings of a Workshop Held in Singapore, May 28–30, 1980*. IDRC, Canada.

Siebert, S. F. 1990. "Impacts of the Rattan Ban on U.S. Business and Some Thoughts on How to Involve Rattan-Importing Businesses in Conservation Efforts." *RIC Bulletin* 9(1):13–15.

Siebert, S. F., and J. M. Belsky. 1985. "Forest Product Trade in a Lowland Filipino Village." *Economic Botany* 39(4):522–33.

Smith, N. 1981. *Man, Fishes and the Amazon*. Columbia University Press, New York, NY.

Tewari, D. N. 1982. *Minor Forest Products of India*. USAID Conference on Forestry and Development in Asia.

United States Department of Agriculture (USDA). 1976. *Tree Nuts*. USDA Foreign Agriculture Circular. July:5–76.

Watt, G. 1889–93. *A Dictionary of the Economic Products of India*. Six volumes. Index by E. Thurston and T. N. Mukerji. Superintendent of Government Printing, India, Calcutta.

Weinstock, J. A. 1983. "Rattan: Ecological Balance in a Borneo Rainforest Swidden." *Economic Botany* 37(1):58–68.

_____. 1985. "Alternate Cycle Agroforestry." *Agroforestry Systems* 3:387–97.

Weinstock, J. A., and N. T. Vergara. 1987. "Land or Plants: Agricultural Tenure in Agroforestry Systems." *Economic Botany* 41(2):312–22.

Westphal, E., and P. C. M. Jansen (eds). 1989. *Plant Resources of Southeast Asia: A Selection*. Volume 1 in a series. Pudoc Wageningen, the Netherlands.

Whitmore, T. C. 1984. *Tropical Rain Forests of the Far East* (2d edition). Clarendon Press, Oxford.

World Health Organization of the United Nations (WHO). 1977. "The Selection of Essential Drugs." Second Report of the WHO Expert Committee. *WHO Technical Report Series* 641.

World Resources Institute (WRI). 1990. *World Resources 1990–1991*. In collaboration with The United Nations Environment Programme and The United Nations Development Programme. Oxford University Press, New York, NY.

Wyatt-Smith, J. 1963. "Manual of Malaysian Silviculture for Inland Forests." *Malaysian Forest* Record No. 23. Forestry Department, Kuala Lumpur, Malaysia.

Yap, S. K. 1982. "The Phenology of Some Fruit Tree Species in a Lowland Dipterocarp Forest." *Malaysian Forester* 45(1):21–35.

CHAPTER SIX

Campbell, J. 1987. "Tropical Forestry and Biological Diversity in India and the Role of U.S. Agency for International Development/New Delhi." August 1987. USAID/New Delhi.

Carpenter, R. (ed.). 1983. *Natural Systems for Development: What Planners Need to Know*. East-West Environment and Policy Institute. Macmillan Publishing, New York, NY.

Caufield, C. 1985. *In the Rainforest: Report from a Strange, Beautiful, Imperiled World*. Alfred A. Knopf, New York, NY.

Center for Science and Environment. 1982. *State of India's Environment—1982: A Citizens Report*. New Delhi, India.

Clay, J. 1982. "Deforestation: The Human Costs." *Cultural Survival Quarterly* 6(2):3.

Daryadi, L. 1981. *Forestry and Agriculture: A Key Connection in Java*. Department of Forestry, Jakarta, Indonesia.

Dickinson, R. E. 1981. "Effects of Tropical Deforestation on Climate." *Studies in Third World Societies* (14):411–42.

Dosso, H., et al. 1981. "The Tai Project: Land Use Problems in a Tropical Forest." *Ambio* 10(2–3):120–25.

Durst, Patrick. 1987. "Natural History and Nature-oriented Adventure Travel for Rural Development and Wildland Management: Diagnosis of Research Needs and Project Opportunities for the Philippines." Forestry Private Enterprise Initiative Working Paper, No. 11. Southeastern Center for Forest Economics, Research Triangle Park, NC.

Fleming, W. M. 1979. *Environmental and Economic Impacts of Watershed Conservation on a Major Reservoir Project in Ecuador*. Environmental Improvement Division, Sante Fe, NM.

Gilmour, D. A. 1977. "Logging and the Environment with Particular Reference to Soil and Stream Protection in Tropical Rain Forest Situations." In *Guidelines for Watershed Management*. FAO Conservation Guide 1, FAO, Rome.

Goulding, M. 1980. *The Fishes and the Forest: Explorations in Amazonian Natural History*. University of California Press, Berkeley, CA.

Grainger, A. 1980. "The State of the World's Tropical Forests." *The Ecologist* 10(1).

Hamilton, L., with P. King. 1983. *Tropical Forested Watersheds: Hydrologic and Soils Response to Major Uses or Conversions*. Westview Press, Boulder, CO.

Hamilton, L., and S. C. Snedaker (eds.). 1984. *Handbook for Mangrove Area Management*. East-West Center, Honolulu, HI.

Hewlitt, J. D. 1982. "Forests and Floods in the Light of Recent Investigation." Proceedings of the Canadian Hydrological Symposium, June 14–15. Fredericton, NB.

Holmes, J. W., and E. B. Wronski. 1982. "On the Water Harvest from Afforested Catchments." Proceedings of the First National Symposium on Forest Hydrology, Melbourne.

Houghton, R. A., et al. 1983. "Changes in the Carbon Content of Terrestrial Biota and Soils Between 1860 and 1980: A Net Release of Carbon Dioxide to the Atmosphere." *Ecological Monographs*.

International Monetary Fund (IMF). 1987. *Primary Commodities: Market Developments and Outlook*. International Monetary Fund, Washington, DC.

International Union for the Conservation of Nature: Natural Resources (IUCN). 1980a. *World Conservation Strategy*. International Union for the Conservation of Nature and Natural Resources. Gland, Switzerland.

———. 1980b. *Averting a Major Setback in Development*. IUCN Bulletin 11(5):122–26.

Jacks, G. V., and R. O. Whyte. 1939. *The Rape of the Earth: A World Survey of Soil Erosion*. Faber and Faber, London.

JRB Associates. 1981. *Dominican Republic: Country Environmental Profile*. McLean, VA.

Keller, M., J. Jacob, S. C. Wofsky, and R. C. Harris. 1990. "Effects of Tropical Deforestation on Global and Regional Atmospherical Chemistry." *Climatic Change* 19:139–58.

Krutilla, J. V. 1991. "Environmental Resource Services of Malaysian Moist Tropical Forests." Mimeograph.

Laarman, J. 1987. "Tropical Science on Economic Activity: OTS in Costa Rica." Draft Manuscript.

Ledec, G. 1985. "The Political Economy of Tropical Deforestation." In H. J. Leonard (ed.), *Divesting Nature's Capital.* John Wiley and Sons, New York, NY.

Leighton, N., and N. Wirawan. 1986. "Catastrophic Drought and Fire in Borneo Tropical Rainforests Associated with the 1983 El Niño Southern Oscillation Event." In G. T. Prance (ed.), *Tropical Forests and the World Atmosphere*, 75–102. American Association for the Advancement of Science (AAAS) Symposium, Washington, DC.

Leonard, H. J. 1986. *Natural Resources and Economic Development in Central America: A Regional Environmental Profile.* Draft Manuscript. International Institute for Environment and Development.

Marsh, G. 1864. *Man and Nature.* Scribner, New York, NY.

McDonald, R., and M. Reisner. 1986. "The High Costs of High Dams." In Maguire and Brown (eds.), *Bordering on Trouble: Resources and Politics in Latin America.* Adler and Adler, Bethesda, MD.

Megahan, W. F. 1977. "Reading Erosional Impacts of Roads." In *Guidelines for Watershed Management.* FAO Conservative Guide 1, Rome.

Megahan, W. F., and W. J. Kidd. 1972. "Effects of Logging and Logging Roads on Erosion and Sediment Deposition from Steep Terrain." *Forestry* 70.

Morgan, R. P. C. 1979. *Soil Erosion.* Longman, London.

Myers, N. 1984. *The Primary Source: Tropical Forests and Our Future.* W.W. Norton and Company, New York and London.

National Academy of Science (NAS). 1982. *Ecological Aspects of Development in the Humid Tropics.* National Academy of Sciences, Washington, DC.

Niehaus, F. 1979. International Atomic Energy Agency *IAEA Bulletin 21* (1):2–10.

Nordhaus, W. D. 1990. "To Curb or Not to Curb: The Economics of the Greenhouse Effect." Paper presented at the Annual Meeting of the American Association for the Advancement of Science, New Orleans, LA.

Office of Technology Assessment (OTA). 1987. *Technologies to Maintain Biological Diversity.* U.S. Government Printing Office, Washington, DC.

Panayotou, T. 1992. "Protecting Tropical Forests." Paper presented at the annual meeting of the American Economic Association, New Orleans, LA.

Pereira, H. C. 1973. *Land Use and Water Resources in Temperate and Tropical Climates.* Cambridge University Press, Cambridge, MA.

Potter, G. L. 1981. "Albedo Change by Man: Test of Climatic Effects." *Nature* 291:47–50.

Potter, G. L., et al. 1975. "Possible Climatic Impact of Tropical Deforestation." *Nature* 258:697–98.

Prescott-Allen, C., and R. Prescott-Allen. 1986. *The First Resource: Wild Species in the North American Economy.* Yale University Press, New Haven, CT.

Ranganathan, S. 1978. "India's Billion Dollar Greenbacks." In N. Myers, *The Primary Source: Tropical Forests and Our Future.* W.W. Norton and Company, New York and London.

Rockefeller Foundation. 1981. *Agricultura de Ladera en America Latina Tropical.* Rockefeller Foundation, New York, NY.

Sagan, C., O. B. Toon, and J. B. Pollack. 1979. "Anthropogenic Albedo Changes and the Earth's Climate." *Science* 206:1363–68.

Salati, E. 1981. "Floresta e Clima da Amazonia." *Bull. Fed. Bras. Conserv. Nat.* 16, Rio de Janeiro.

Soemarwoto, O. 1979. *Ecological and Environmental Impacts of Energy Use in Asian Developing Countries, with Particular Reference to Indonesia.* Institute of Ecology, Bandung, Indonesia.

Soulé, M. E. 1982. Address to Earthscan Seminar on Genetic Resources. Bali, October 1982.

Spears, J. 1982. "Preserving Wasted Environments." *Unasylva* 34.

Stone, R. 1985. *Dreams of Amazonia.* Viking Penguin, New York, NY.

Sumitro, A. (ed.). 1979. *Population and Environment Planning for Communities Practicing Shifting Cultivation.* Gadjah Mada University, Yogyakarta, Indonesia.

Swank, W. T., and J. E. Douglass. 1974. "Streamflow Greatly Reduced by Converting Deciduous Hardwood Stands to Pine." *Science* 185:857–59.

Swank, W. T., and N. H. Miner. 1968. "Conversion of Hardwood-Covered Watersheds to White Pine Reduces Water Yield." *Water Resources Research* 4:947–54.

Timberlake, L. 1987. *Only One Earth.* BBC Books and Earthscan, London.

United States Fish and Wildlife Service. 1981. *1980 National Survey of Fishing, Hunting, and Wildlife-Associated Recreation.* U.S. Government Printing Office, Washington, DC.

Vietmeyer, N. 1975. *Underexploited Tropical Plants with Promising Economic Value.* National Academy of Science, Washington, DC.

Weber, W., and A. Vedder. 1984. "Forest Conservation in Rwanda and Burundi." *Swara.* East African Wildlife Society. 7:32–36.

Whitmore, T. C. 1984. *Tropical Rain Forests of the Far East* (2d edition). Clarendon Press, Oxford.

Wiersum, K. F. 1984. "Surface Erosion under Various Tropical Agroforestry Systems." In C. L. O'Loughlin and A. J. Pearce (eds.), *Symposium on Effects of Forest Land Use on Erosion and Slope Stability*, 231–39. Environment and Policy Institute, East-West Center, Honolulu, HI.

Wijkman, A., and L. Timberlake. 1984. *Natural Disasters: Acts of God or Acts of Man?* BBC Books and Earthscan, International Institute for Environment and Development, London and Washington, DC.

Williams, J., and L. Hamilton. 1982. *Watershed Forest Influences in the Tropics and Subtropics: A Selected, Annotated Bibliography.* East-West Center, Honolulu, HI.

Woodwell, G. M. 1978. *Scientific American* 238(1):34–43.

Woodwell, G. M., et al. 1983. "Global Deforestation: Contribution to Atmospheric Carbon Dioxide." *Science* 222:1081–86.

World Bank. 1987. *Wildlands: Their Protection and Management in Economic Development.* Washington, DC.

————— 1991. *The Forest Sector: A World Bank Policy Paper*. Washington, DC.

World Wildlife Fund (WWF). 1980. *Strategy for Training in Natural Resources and Environment: A Proposal for the Development of Personnel and Institutions in Latin America and the Caribbean*. Washington, DC.

CHAPTER SEVEN

Anderson, D. 1987. *The Economics of Afforestation: A Case Study in Africa*. World Bank Occasional Paper Number 1. The Johns Hopkins University Press, Baltimore and London.

Briscoe, J., P. Furtado de Castro, C. Griffin, J. North, and O. Olsen. 1990. "Toward Equitable and Sustainable Rural Water Supplies: A Contingent Valuation Study in Brazil." *The World Bank Economic Review* 4(2):115–34.

Clawson, M. 1974. "Economic Trade-offs in Multiple Use Management of Forest Lands." *American Journal of Agricultural Economics*, December.

Dixon, J. A., and M. M. Hufschmidt (eds.). 1986. *Economic Valuation Techniques for the Environment*. Johns Hopkins University Press, Baltimore, MD.

Godoy, R., and R. Lubowski. 1991. "How Much Is the Forest Worth? Guidelines for the Valuation of Non-Timber Tropical Forest Products." Unpublished Manuscript. Harvard Institute for International Development, Cambridge, MA.

Gregersen, H. 1982. "Valuing Goods and Services from Tropical Forests and Woodlands." Unpublished Manuscript.

Hartwick, J., and N. Olewiler. 1986. *The Economics of Natural Resources Use*. Harper & Row Publishers, Inc., New York, NY.

Hufschmidt, M. M., D. E. James, A. D. Meister, B. T. Bower, and J. A. Dixon. 1983. *Environment, Natural Systems, and Development*. The Johns Hopkins University Press, Baltimore, MD.

Kneese and J. Sweeney (eds.). 1985. *Handbook of Natural Resource and Energy Economics*, Vol. 2. North-Hollard, New York, NY.

Mitchell, R. C., and R. T. Carson. 1989. *Using Surveys to Value Public Goods: The Contingent Valuation Method*. Resources for the Future, Washington, DC.

Myers, N. 1984. *The Primary Source: Tropical Forests and Our Future*. W.W. Norton & Company, New York, NY.

Peters, C. M., A. H. Gentry, and R. Mendelsohn. 1989. "Valuation of a Tropical Forest in Peruvian Amazonia." *Nature* 339:655–56.

Schmink, Marianne. 1986. "The Rationality of Tropical Forest Destruction." USDA Forestry Service, Puerto Rico.

Sfeir-Younis, Alfredo. 1987. *Methods for Appropriate Economic Valuation of Natural Resources in Developing Countries*. Unpublished Manuscript.

Smith, N. 1981. *Man, Fishes and the Amazon*. Columbia University Press, New York, NY.

Vincent, J. R., E. W. Crawford, and J. P. Hoehn (eds.). 1991. *Valuing environmental benefits in developing economies*. Special Report 29, Michigan Agricultural Experiment Station, Michigan State University, East Lansing, MI.

Whittington, D., D. T. Lauria, and X. Mu. 1989. "A Study of Water Vending and Willingness to Pay for Water in Onitsha, Nigeria." Unpublished Manuscript.

Whittington, D., D. T. Lauria, A. M. Wright, K. Choe, J. A. Hughes, and V. Swarna. 1991. "Willingness to Pay for Improved Sanitation in Kumasi, Ghana: A Contingent Valuation Study." A report to the Infrastructure and Urban Development Department, the World Bank. Draft Manuscript.

CHAPTER EIGHT

Appanah, S., and Salleh Mohd. Nor. 1987. "Natural Regeneration and Its Implications for Forest Management in the Malaysian Dipterocarp Forests." Draft Manuscript, July.

Asabere, P. K. 1987. "Attempts at Sustained Yield Management in the Tropical High Forests of Ghana." In F. Mergen and J. R. Vincent (eds.), *Natural Management of Tropical Moist Forest*, 47–70. Yale University, School of Forestry and Environmental Studies, New Haven, CT.

Caldecott, J. 1987. *Hunting and Wildlife Management in Sarawak*. World Wildlife Fund, Malaysia.

Ewel, J. 1981. "Environmental Implications of Tropical Forest Utilization." In F. Mergen (ed.), *International Symposium on Tropical Forest Utilization and Conservation*, 156–67. Yale University Press, New Haven, CT.

Fox, J. E. D. 1968. "Logging Damage and Influence of Climber Cutting Prior to Logging in the Lowland Dipterocarp Forest of Sabah." *Malaysian Forester*. XXXI:326–47.

Hamilton, L., with P. King. 1983. *Tropical Forested Watersheds*. Westview Press, Inc., Boulder, CO.

Hutchinson, I. D. 1986. Unpublished Manuscript. Institute of Tropical Forestry, USDA Forest Service.

Johns, A. 1983. "Wildlife Can Live with Logging." *New Scientist* 6(21):206–11.

Kio, P. R. O. 1987. "ITTO Project: Perspective for Africa." Consultancy Report for the Harvard Institute for International Development, Cambridge, MA.

Kio, P. R. O., and S. A. Ekwebelam. 1987. "Plantations versus Natural Forests for Meeting Nigeria's Wood Needs." In F. Mergen and J. R. Vincent (eds.), *Natural Management of Tropical Moist Forest*, 149–76. Yale University, School of Forestry and Environmental Studies, New Haven, CT.

Leighton, M., and D. R. Leighton. 1983. "Vertebrate Responses to Fruiting Seasonality within a Bornean Rain Forest." In Sutton, Whitmore, and Chadwick (eds.), *Tropical Rain Forest Ecology and Management*. Blackwell, Oxford.

Liew That Chim and Wong Fung On. 1973. "Density, Recruitment, Mortality and Growth of Dipterocarp Seedlings in Virgin and Logged-over Forests in Sabah." *Malaysian Forester* XXXVI(1):3–15.

Marn, H. M., and W. B. Jonkers. 1981. "Logging Damage in Tropical High Forest." Paper presented at the International Forestry Seminar, Kuala Lumpur. Sarawak Forestry Department, Kuching, Sarawak, Malaysia.

McClure, H. E. 1966. "Flowering, Fruiting and Animals in the Canopy of a Tropical Rain Forest." *Malaysian Forester* XXIX:182–203.

Nicholson, D. I. 1958. "An Analysis of Logging Damage in Tropical Rain Forest, North Borneo." *Malaysian Forester* XXI:235–45.

Nwoboshi, C. L. 1987. "Regeneration Success of Natural Management, Enrichment Planting, and Plantations of Native Species in West Africa." In F. Mergen and J. R. Vincent (eds.), *Natural Management of Tropical Moist Forest*, 71–92. Yale University, School of Forestry and Environmental Studies, New Haven, CT.

Office of Technology Assessment (OTA), United States Congress. 1984. *Technologies to Sustain Tropical Forest Resources*. U.S. Government Printing Office, Washington, DC.

Schmidt, R. 1987. "Tropical Rain Forest Management: A Status Report." *Unasylva* 39(156):2–17.

Spears, J. 1987. "Indonesia: Forestry Appraisal Mission; Working Paper on Economic Rates of Return for Various Natural Forest Management/Plantation Options." Unpublished Manuscript.

Whitmore, T. C. 1984. *Tropical Rain Forests of the Far East* (2d edition). Clarendon Press, Oxford.

CHAPTER NINE

Allegretti, M. H., and S. Schwartzman. 1986. "Extractive Resources: A Sustainable Development Alternative for Amazonia." Mimeograph Report to the World Wildlife Fund. WWF, Washington, DC.

Asabere, P. K. 1987. "Attempts at Sustained Yield Management in the Tropical High Forests of Ghana." In F. Mergen and J. R. Vincent (eds.), *Natural Management of Tropical Moist Forest*, 47–70. Yale University, School of Forestry and Environmental Studies, New Haven, CT.

Ashton, P. S. In Prep. "Ecological Studies in the Mixed Dipterocarp Forests of Northwest Borneo: Floristic Variation and Patterns of Species Richness." Harvard University, Cambridge, MA.

Ashton, P. M. S., and P. S. Ashton. 1992. "Plant Resources for Agroforestry Systems in South and Southeast Asia." In W. R. Bentley, P. Khosla, and K. Seckler (eds.), *Agroforestry Systems of South Asia*, pp. 56–75. Oxford IBH, New Delhi, India.

Ashton, P. M. S., I. A. U. N. Gunatilleke, and C. V. S. Gunatilleke. In press. "A Case for the Development and Evaluation of Mixed-species Plantations in Southwest Sri Lanka." In E. Erdelen, Ch. Preu, N. Ishawaran, and Ch. Santiapillai (eds.), *Ecology and Landscape Management in Sri Lanka; Conflict or Compromise?* Josef Margraf Scientific, Hamburg, Germany.

Ashton, P. S., and P. Hall. 1992. "Comparisons of Structure among Mixed Dipterocarp Forests of Northwestern Borneo." *Journal of Ecology* 80:(pages not yet available).

Augspurger, C. K., and C. K. Kelly. 1984. "Pathogen Mortality of Tropical

Tree Seedlings: Experimental Studies of the Effects of Dispersal Distance, Seedling Density, and Light Conditions." *Oecologia* 61:211–17.

Bromley, D. W. 1981. "The Economics of Social Forestry: An Analysis of a Proposed Program in Madhya Pradesh, India." Unpublished Report to USAID, Center for Resource Policy Studies, University of Wisconsin, Madison, WI.

Carpenter, R. A. 1983. *Natural Systems for Development: What Planners Need to Know*. Macmillan Publishing Company, New York, NY.

Christanty, L., M. Hadyana, Sigit, and Projono. 1980. "Distribution and Interception of Light in a Homegarden." Unpublished paper cited in O. Soemarwoto and I. Soemarwoto, "The Javanese Rural Ecosystem." In T. A. Rambo and P. E. Sajise (eds.), *An Introduction to Human Ecology Research on Agricultural Systems in Southeast Asia*. East-West Environment Policy Institute and University of the Philippines at Los Banos, Philippines.

Clark, D. A., and D. B. Clark. 1984. "Sapling Dynamics of a Tropical Rainforest Tree: Evaluation of the Janzen-Connell Model." *American Naturalist* 124:769–88.

Danoesastro, H. 1980. "The Role of Homegarden as a Source of Additional Family Income." Unpublished Paper cited in O. Soemarwoto and I. Soemarwoto, "The Javanese Rural Ecosystem." In T. A. Rambo and P. E. Sajise (eds.), *An Introduction to Human Ecology Research on Agricultural Systems in Southeast Asia*. East-West Environment Policy Institute and University of the Philippines at Los Banos, Philippines.

Evans, J. 1982. *Plantation Forestry in the Tropics*. Oxford University Press, Oxford.

Godoy, R. 1988. "Prospects for Rattan Cultivation: Preliminary Assessment." Memorandum to the Government of Indonesia.

Gomez-Pompa, A. 1987. "On Maya Silviculture." *Mexican Studies/Estudios Mexicanos* 3(1):1–17.

Grainger, A. 1986. "The Future Role of the Tropical Rain Forests in the World Forest Economy." Ph.D. Dissertation, Oxford University, St. Cross College.

————. 1988. "Future Supplies of High-Grade Tropical Hardwoods from Intensive Plantations." *Journal of World Forest Resource Management* 3:15–29.

Hartshorn, G. 1983. "Ecological Implications of Tropical Plantation Forestry." In R. Sedjo, *The Comparative Economics of Plantation Forestry: A Global Assessment*. Resources for the Future, Inc., Johns Hopkins University Press, Baltimore, MD.

Hubbell, S. P. 1980. "Seed Predation and the Coexistence of Tree Species in Tropical Forests." *Oikos* 35:214–29.

Hubbell, S. P., and R. B. Foster. 1987. "Structure, Dynamics and Equilibrium Status of Old-growth Forest on Barro Colorado Island." In A. L. Gentry (ed.), *Four Tropical Forests*, 522–41. Yale University Press, New Haven, CT.

Janzen, D. H. 1970. "Herbivory and the Number of Tree Species in Tropical Forests." *American Naturalist* 104:501–28.

Karyono. 1981. "Homegarden Structure in the Villages of the Citarum River Basin, West Java." Doctoral Dissertation, Padjadjaran University, Bandung, Indonesia (Indonesian, mimeographed). In O. Soemarwoto and I. Soemarwoto, "The Javanese Rural Ecosystem" in T. A. Rambo and P. E. Sajise, *An Introduction to Human Ecology Research on Agricultural Systems in Southeast Asia*, 254–87. East-West Environment Policy Institute and University of the Philippines at Los Banos, Philippines.

Kio, P. R. O. 1987. "Perspective for Africa." Unpublished Consultancy Report to the Harvard Institute for International Development (HIID). Cambridge, MA.

Lanly, J. 1982. *Tropical Forest Resources*. FAO Forestry Paper No. 30. FAO, Rome.

Leslie, A. 1977. "Where Contradictory Theory and Practice Co-exist." *Unasylva* 29(115):2–17.

Lowe, R. G., and F. N. Ugbechie. 1975. "Kennedy's Natural Regeneration Experiments at Sapoba, After 45 Years." Unpublished Report to the Forestry Association of Nigeria, 6th Annual Conference. Calabar, Nigeria.

McGaughey, S. E., and H. M. Gregersen (eds.). 1983. *Forest Based Development in Latin America*. Inter-American Development Bank, Washington, DC.

Moad, A. S. 1986. "Segaluid-Lokan Forest Plantation Project: Mixed Dipterocarp Plantation Experiments." Unpublished Report to the Sabah Forestry Department. Sandakan, Malaysia.

National Research Council (NRC). 1982. *Ecological Aspects of Development in the Humid Tropics*. National Academy Press, Washington, DC.

Raintree, J. B. 1986. "Agroforestry Pathways: Land Tenure, Shifting Cultivation and Sustainable Agriculture." *Unasylva* 38(154):2–15.

Sedjo, R. 1983. *The Comparative Economics of Plantation Forestry: A Global Assessment*. Resources for the Future, Johns Hopkins University Press, Baltimore, MD.

————. 1987. "The Economics of Natural and Plantation Forests in Indonesia." Unpublished Manuscript.

Soemarwoto, O., and I. Soemarwoto. 1984. "The Javanese Rural Ecosystem." In T. A. Rambo and P. E. Sajise (eds.), *An Introduction to Human Ecology Research on Agricultural Systems in Southeast Asia*. East-West Environment Policy Institute and University of the Philippines at Los Banos, Philippines.

Spears, J. 1979. "Can the Wet Tropical Forest Survive?" *Commonwealth Forestry Review* 58(3):165–80.

————. 1982. "Preserving Watershed Environments." *Unasylva* 34(137):10–14.

————. 1987. "Working Paper on Economic Rates of Return for Various Natural Forest Management/Plantation Options." Unpublished Manuscript. Forestry Appraisal Mission, Indonesia.

Spears, J., and E. S. Ayensu. 1985. "Resources, Development, and the New Century: Forestry." In R. Repetto (ed.), *The Global Possible: Resources, Development, and the New Century*. Yale University, Vail-Ballou Press, Birmingham, NY.

Tan, S. S., R. Primack, E. O. K. Chai, and H. S. Lee. 1987. "The Silviculture of Dipterocarp Trees in Sarawak, East Malaysia, III: Plantation Forest." *Malaysian Forester* 50:148–61.

United Nations Educational, Scientific, and Cultural Organization (UNESCO). 1978. *Tropical Forest Ecosystems: A State of Knowledge Report*. UNESCO, Presses Université de France, Paris.

Whitmore, T. C. 1984. *Tropical Rain Forests of the Far East* (2d. edition). Clarendon Press, Oxford.
World Resources Institute (WRI). 1985. *Tropical Forests: A Call for Action (Parts 1
and 2)*. World Resources Institute, Washington, DC.

CHAPTER TEN

Ashton, P. S. 1967. "Climate versus Soils in the Classification of Southeast
Asian Tropical Lowland Vegetation." *Journal of Ecology* 55:67.
_____. 1977. "A Contribution of Rain Forest Research to Evolutionary The-
ory." *Annals of the Missouri Botanical Garden* 64:696–705.
_____. 1981. "Techniques for the Identification and Conservation of
Threatened Species in Tropical Forests." In H. Synge (ed.), *The Biological
Aspects of Rare Plant Conservation*. Wiley, Chichester.
_____. 1984. "Biosystematics of Tropical Forest Plants: A Problem of Rare
Species." In W. Grant (ed.), *Plant Biosystematics*. Academic Press, Toronto.
_____. In Prep. "Ecological Studies in the Mixed Dipterocarp Forests of
Northwest Borneo: Floristic Variation and Patterns of Species Richness."
Harvard University, Cambridge, MA.
Ashton, P. S., and P. Hall. 1992. "Comparisons of Structure among Mixed
Dipterocarp Forests of Northwestern Borneo." *Journal of Ecology* 80:(pages
not yet available).
Bawa, K. S. 1976. "Breeding Systems of Tree Species of a Lowland Tropical
Community." *Evolution* 28:85–92.
Bowes, M. D., and J. V. Krutilla. 1989. *Multiple-Use Management: The Economics
of Public Forestlands*. Resources for the Future, Washington, DC.
Caldecott, J. 1987. *Hunting and Wildlife Management in Sarawak*. World Wildlife
Fund, Malaysia.
Diamond, J. A., and T. J. Case. 1986. *Community Ecology*. Harper & Row, New
York, NY.
Forman, R. T. T., and M. Godran. 1986. *Landscape Ecology*. J. Wiley, New
York, NY.
Gan, Y. Y., F. W. Robertson, and E. Soepadmo. 1981. "Isozyme Variations in
Some Rain Forest Trees." *Biotropica* 13:20–28.
Gentry, A. M. 1988. "Changes in Plant Community Diversity and Floristic
Composition on Environmental and Geographical Gradients." *The Annals of
the Missouri Botanical Garden* 75:1–34.
Hamrick, J. L. 1983. "The Distribution of Genetic Variation within and among
Natural Plant Populations." In C. M. Schonewald-Cox, S. M. Cham-
bers, B. MacBryde, and W. L. Thomas (eds.), *Genetics and Conservation*.
Benjamin Cummings, Menlo Park, CA.
Hamrick, J. L., and D. A. Murawski. 1991. "Levels of Allozyme Diversity in
Populations of Uncommon Neotropical Tree Species." *Journal of Tropical
Ecology* 5:157–65.
Hubbell, S. P., and R. B. Foster. 1987. "Structure, Dynamics and Equilibrium

Status of Old-growth Forest on Barro Colorado Island." In A. L. Gentry, *Four Neotropical Forests*, 522–41. Yale University Press, New Haven, CT.

Huston, M. 1978. "A General Hypothesis of Species Diversity." *American Naturalist* 113:81–101.

Janzen, D. H. 1971. "Seed Predation by Animals." *Annual Review of Ecol. System* 2:465–92.

Kenworthy, J. B. 1971. "Water and Nutrient Cycling in a Tropical Rain Forest." In J. R. Flenley (ed.), *The Water Relations of Malaysian Forests* 49–65. Department of Geography, University of Hull.

Ledig, F. T. 1986. "Heterozygosity, Heterosis, and Fitness in Outbreeding Plants." In M. E. Soulé (ed.), *Conservation Biology: The Science of Scarcity and Diversity*, 77–104. Sinauer, Sunderland, MA.

Lovejoy, T. E., R. O. Bierregard, Jr., A. S. Rylands, J. R. Malcolm, C. E. Quintela, L. H. Harper, K. S. Brown, Jr., A. H. Powell, G. V. N. Powell, H. O. R. Schubart, and M. B. Hayes. 1986. "Edge and Other Effects of Isolation on Amazon Forest Fragments." In M. E. Soulé (ed.), *Conservation Biology: The Science of Scarcity and Diversity*, 257–85. Sinauer, Sunderland, MA.

Macarthur, R. H., and E. O. Wilson. 1967. *The Theory of Island Biogeography*. Princeton Monographs in Population Biology I. Princeton University Press, Princeton, NJ.

Mayr, E. 1963. *Animal Species and Evolution*. Harvard University Press, Cambridge, MA.

Murawski, D. A., and J. L. Hamrick. 1991. "The Effect of Density of Flowering Individuals on the Mating Systems of Nine Tropical Tree Species." *Heredity* 67:167–74.

Raven, P. R. (ed.). 1980. *Research Priorities in Tropical Biology*. National Research Council, Washington, DC.

Ridley, H. N. 1930. *The Dispersal of Plants throughout the World*. L. Reeves, Ashford, England.

Tilman, D. 1982. *Resource Competition and Community Structure*. Princeton Monographs in Population Biology. Princeton University Press, Princeton, NJ.

Whitmore, T. C. 1977. "A First Look at Agathis." *Tropical Forestry Papers II*. Oxford Forestry Institute, Oxford.

World Bank. 1987. *Wildlands: Their Protection and Management in Economic Development*. World Bank, Washington, DC.

CHAPTER ELEVEN

Cernea, M. 1981. *Land Tenure Systems and Social Implications of Forestry Development Programs*. World Bank Staff Working Paper No. 452. The World Bank, Washington, DC.

Dasgupta, P. 1982. *The Control of Resources*. Blackwell, Oxford.

Feder, Gershon, T. Onchan, and C. Hongladarom. 1986. *Land Ownership Security, Farm Productivity, and Land Policies in Rural Thailand*. Research Project RPO 673–33. World Bank, Washington, DC.

Food and Agricultural Organization of the United Nations (FAO). 1981. *Forest Resources of Tropical Asia*. Tropical Forest Resources Assessment Project (in the framework of the Global Environment Monitoring System—GEMS). FAO, Rome.

Hardin, G. 1968. "The Tragedy of the Commons." *Science* 162:1243–48.

Lanly, J. 1982. *Tropical Forest Resources*. Forestry Paper No. 30. FAO, Rome.

Repetto, R. 1986. *Economic Policy Reform for Natural Resource Conservation*. World Resources Institute Draft Paper. Washington, DC.

Repetto, R., and M. Gillis (eds.). 1988. *Public Policies and the Misuse of Forest Resources*. Cambridge University Press, New York, NY.

Thailand Development Resources Institute (TDRI). 1987. *Thailand Natural Resources Profile: Is the Resource Base for Thailand's Development Sustainable?* Bangkok, Thailand.

CHAPTER TWELVE

Gillis, M. 1987. "Deforestation and the Tropical Asian Wood Products Industry." In *People of the Tropical Rain Forest*. Smithsonian Institution, Washington, DC.

———. 1988. "Indonesia: Public Policies, Resource Management, and the Tropical Forest." In R. Repetto and M. Gillis (eds.), *Public Policies and the Misuse of Forest Resources*, 43–113. Cambridge University Press, New York, NY.

Lennertz, R., and K. F. Panzer. 1983. "East Kalimantan Transmigration: Development Project PN 76.2010.7 Report." Unpublished. Referred to in N. Leighton and N. Wirawan, "Catastrophic Drought and Fire in Borneo Tropical Rainforests Associated with the 1983 El Niño Southern Oscillation Event." In G. T. Prance (ed.), *Tropical Forests and the World Atmosphere*, 75–102. American Association for the Advancement of Science Symposium, Washington, DC.

Leslie, A. J. 1987. "A Second Look at the Economics of Natural Forest Management Systems in Tropical Mixed Forests." *Unaslyva* 39(155).

McGaughey, S. E., and H. M. Gregersen (eds.). 1983. *Forest Based Development in Latin America*. InterAmerican Development Bank, Washington, DC.

Panayotou, T. 1990. "Investing in Rural Industry for Employment and Balanced Growth: The Case of Thailand." In P. N. Nemetz (ed.), *The Pacific Rim: Investment, Development and Trade*. University of British Columbia Press, Vancouver.

———. 1991. "Economic Incentives in Environmental Management and Their Relevance to Developing Countries." In D. Eröcal (ed.), *Environmental Management in Developing Countries*. Organization of Economic Cooperation and Development (OECD), Paris.

Repetto, R. 1986. "Economic Policy Reform for Natural Resource Conservation." Draft Report. World Resources Institute, Washington, DC.

World Bank. 1987. *Wildlands: Their Protection and Management in Economic Development*. World Bank, Washington, DC.

CHAPTER THIRTEEN

McGaughey, S. E., and H. M. Gregerson (eds.). 1983. *Forest Based Development in Latin America.* InterAmerican Development Bank, Washington, DC.

CHAPTER FOURTEEN

Anonymous. 1989. *Funding Priorities for Research toward Effective Sustainable Management of Biodiversity Resources in Tropical Asia.* Report of a workshop sponsored by the National Science Foundation and the United States Agency for International Development, Bangkok, March 27–30.

Manokaran, N., J. V. LaFrankie, K. M. Kochummen, E. S. Quah, J. E. Klahn, P. S. Ashton, and S. P. Hubbell. 1990. "Methodology for the Fifty-Hectare Research Plot at Pasoh Forest Reserve." *Forest Research Institute, Malaysia Research Pamphlet 104.*

Panayotou, T. 1992. "Protecting Tropical Forests." Paper presented at the annual meeting of the American Economic Association, New Orleans, LA.

Index

About the Authors

THEODORE PANAYOTOU is a fellow of the Harvard Institute of International Development (HIID) and a lecturer in the Department of Economics at Harvard University. He is also a member of the Center for Tropical Forest Science. A specialist in environmental and resource economics, environmental policy analysis, and development economics, Dr. Panayotou has advised governments and institutes in Asia, Africa, and Eastern Europe as well as numerous other national and international institutions, on the interactions between the natural resource base and economic development. Dr. Panayotou served for a decade as visiting professor and resident advisor in Southeast Asia, and recently coauthored a multi-volume environmental policy study at the Thailand Development Research Institute in Bangkok. The author of several books, and numerous articles and monographs including *Green Markets: the Economics of Sustainable Development*, he is currently at work on another book entitled *Natural Resources, Environment, and Development: Economics, Policy, and Management*. Dr. Panayotou received the 1991 Distinguished Achievement Award of the Society for Conservation Biology for his wide-ranging efforts to use economic analysis as a tool for conservation.

PETER S. ASHTON is a faculty fellow of the Harvard Institute for International Development, and Charles Bullard Professor of Forestry in the Department of Organismic and Evolutionary Biology at Harvard University. He is a founding member of the Center for Tropical Forest Science and the former director of the Arnold Arboretum at Harvard. Dr. Ashton's work has focused on the ecology of the rain forests of Asia and the biology of their trees. He has collaborated in research in Malaysia, Indonesia, the Philippines, Thailand, Cambodia, Vietnam, and Sri Lanka. The author of six books and more than one hundred published articles, Dr. Ashton has written most recently on regeneration in mixed dipterocarp forests on different soils, the influence of nutrient availability on rain forest community structure, and species

279

richness in plant communities. He is the 1987 recipient of the Environmental Merit Award from the United States Environmental Protection
Agency, for outstanding contributions to conservation, for his work as
Director of the Arnold Arboretum and the quality and restoration of
the collection, and for research in future sustainment of tropical rain
forests.

ALSO AVAILABLE FROM ISLAND PRESS

Balancing on the Brink of Extinction: The Endangered Species Act and Lessons for the Future
Edited by Kathryn A. Kohm

Better Trout Habitat: A Guide to Stream Restoration and Management
By Christopher J. Hunter

Beyond 40 Percent: Record-Setting Recycling and Composting Programs
The Institute for Local Self-Reliance

Coastal Alert: Ecosystems, Energy, and Offshore Oil Drilling
By Dwight Holing

The Complete Guide to Environmental Careers
The CEIP Fund

Death in the Marsh
By Tom Harris

Farming in Nature's Image
By Judith D. Soule and Jon K. Piper

The Global Citizen
By Donella H. Meadows

Healthy Homes, Healthy Kids
By Joyce M. Schoemaker and Charity Y. Vitale

Holistic Resource Management
By Allan Savory

Inside the Environmental Movement: Meeting the Leadership Challenge
By Donald Snow

Last Animals at the Zoo: How Mass Extinction Can Be Stopped
By Colin Tudge

Learning to Listen to the Land
Edited by Bill Willers

Lessons from Nature: Learning to Live Sustainably on the Earth
By Daniel D. Chiras

The Living Ocean: Understanding and Protecting Marine Biodiversity
By Boyce Thorne-Miller and John G. Catena

Making Things Happen: How to Be an Effective Volunteer
By Joan Wolfe

Media and the Environment
Edited by Craig L. LaMay and Everette E. Dennis

Nature Tourism: Managing for the Environment
Edited by Tensie Whelan

The New York Environment Book
By Eric A. Goldstein and Mark A. Izeman

Our Country, The Planet: Forging a Partnership for Survival
By Shridath Ramphal

Overtapped Oasis: Reform or Revolution for Western Water
By Marc Reisner and Sarah F. Bates

Plastics: America's Packaging Dilemma
By Nancy Wolf and Ellen Feldman

Race to Save the Tropics: Ecology and Economics for a Sustainable Future
Edited by Robert Goodland

Rain Forest in Your Kitchen: The Hidden Connection Between Extinction and Your Supermarket
By Martin Teitel

The Rising Tide: Global Warming and World Sea Levels
By Lynne T. Edgerton

The Snake River: Window to the West
By Tim Palmer

Steady-State Economics: Second Edition with New Essays
By Herman E. Daly

Taking Out the Trash: A No-Nonsense Guide to Recycling
By Jennifer Carless

Trees, Why Do You Wait? America's Changing Rural Culture
By Richard Critchfield

Turning the Tide: Saving the Chesapeake Bay
By Tom Horton and William M. Eichbaum

War on Waste: Can America Win Its Battle With Garbage?
By Louis Blumberg and Robert Gottlieb

Western Water Made Simple
From *High Country News*

For a complete catalog of Island Press publications, please write: Island Press, Box 7, Covelo, CA 95428, or call: 1-800-828-1302.